ROBOT ETHICS

The MIT Press Essential Knowledge Series

A complete list of books in this series can be found online at https://mitpress.mit.edu/books/series/mit-press-essential-knowledge-series.

ROBOT ETHICS

MARK COECKELBERGH

The MIT Press | Cambridge, Massachusetts | London, England

The MIT Press would like to thank the anonymous peer reviewers who provided comments on drafts of this book. The generous work of academic experts is essential for establishing the authority and quality of our publications. We acknowledge with gratitude the contributions of these otherwise uncredited readers.

This book was set in Chaparral Pro by New Best-set Typesetters Ltd. Printed and bound in the United States of America.

Library of Congress Cataloging-in-Publication Data

Names: Coeckelbergh, Mark, author.
Title: Robot ethics / Mark Coeckelbergh.
Description: Cambridge, Massachusetts : The MIT Press, [2022] | Series: The MIT Press essential knowledge series | Includes bibliographical references and index.
Identifiers: LCCN 2021033925 | ISBN 9780262544092
Subjects: LCSH: Robots—Moral and ethical aspects. | Robots—Social aspects.
Classification: LCC TJ211.28 .C64 2022 | DDC 174/.9629892—dc23
LC record available at https://lccn.loc.gov/2021033925

10 9 8 7 6 5 4 3 2 1

CONTENTS

SERIES FOREWORD

The MIT Press Essential Knowledge series offers accessible, concise, beautifully produced pocket-size books on topics of current interest. Written by leading thinkers, the books in this series deliver expert overviews of subjects that range from the cultural and the historical to the scientific and the technical.

In today's era of instant information gratification, we have ready access to opinions, rationalizations, and superficial descriptions. Much harder to come by is the foundational knowledge that informs a principled understanding of the world. Essential Knowledge books fill that need. Synthesizing specialized subject matter for nonspecialists and engaging critical topics through fundamentals, each of these compact volumes offers readers a point of access to complex ideas.

1

INTRODUCTION: WHAT IS ROBOT ETHICS ABOUT?

In the 2004 US science-fiction film *I, Robot*, humanoid robots serve humanity. Yet not all is going well. After an accident, a man is rescued from the sinking car by a robot, but a twelve-year-old girl is not saved. The robot calculated that the man had a higher chance of survival; humans may have made another choice. Later in the film, robots try to take over power from humans. The robots are controlled by an artificial intelligence (AI), VIKI, which decided that restraining human behavior and killing some humans will ensure the survival of humanity. The film illustrates the fear that humanoid robots and AI are taking over the world. It also points to hypothetical ethical dilemmas should robots and AI reach general intelligence. But is this what robot ethics is and should be about?

Are the Robots Coming or Are They Already Here?

Usually when people think about robots, the first image that comes to mind is a highly intelligent, humanlike robot. Often that image is derived from science fiction, where we find robots that look and behave more or less like humans. Many narratives warn about robots that take over; the fear is that they are no longer our servants but instead make us into *their* slaves. The very term "robot" means "forced labor" in Czech and appears in Karel Čapek's play *R.U.R.*, staged in Prague in 1921—just over a hundred years ago. The play stands in a long history of stories about humanlike rebelling machines, from Mary Shelley's *Frankenstein* to films such as *2001: A Space Odyssey*, *Terminator*, *Blade Runner*, and *I, Robot*. In the public imagination, robots are frequently the object of fear and fascination at the same time. We are afraid that they will take over, but at the same time it is exciting to think about creating an artificial being that is like us. Part of our romantic heritage, robots are projections of our dreams and nightmares about creating an artificial other.[1]

First these robots are mainly scary; they are monsters and uncanny. But at the beginning of the twenty-first century, a different image of robots emerges in the West: the robot as companion, friend, and perhaps even partner. The idea is now that robots should not be confined to industrial factories or remote planets in space. In the contemporary imagination, they are liberated from their dirty slave work, and enter the home as pleasant, helpful, and sometimes sexy social partners you can talk to. In some films, they still ultimately rebel—think about *Ex Machina*, for example—but generally they become what robot designers call "social robots." They are designed for "natural" human-robot interaction—that is, interaction in the way that we are used to interacting with other humans or pets. They are designed to be not scary or monstrous but instead cute, helpful, entertaining, funny, and seductive.

This brings us to real life. The robots are not coming; they are already here. But they are not quite like the robots we meet in science fiction. They are not like Frankenstein's monster or the Terminator. They are industrial robots and, sometimes, "social robots." The latter are not as intelligent as humans or their science-fiction kin, though, and often do not have a human shape. Even sex robots are not as smart or conversationally capable as the robot depicted in *Ex Machina*. In spite of recent developments in AI, most robots are not humanlike in any sense. That said, robots are here, and they are here to stay. They are more intelligent and more capable of autonomous functioning than before. And there are more real-world applications. Robots are not only used in industry but also health care, transportation, and home assistance.

Often this makes the lives of humans easier. Yet there are problems too. Some robots may be dangerous

The robots are not coming; they are already here. But they are not quite like the robots we meet in science fiction.

indeed—not because they will try to kill or seduce you (although "killer drones" and sex robots are also on the menu of robot ethics), but usually for more mundane reasons such as because they may take your job, may deceive you into thinking that they are a person, and can cause accidents when you use them as a taxi. Such fears are not science fiction; they concern the near future. More generally, since the impact of nuclear, digital, and other technologies on our lives and planet, there is a growing awareness and recognition that technologies are making fundamental changes to our lives, societies, and environment, and therefore we better think more, and more critically, about their use and development. There is a sense of urgency: we better understand and evaluate technologies now, before it is too late—that is, before they have impacts nobody wants. This argument can also be made for the development and use of robotics: let us consider the ethical issues raised by robots and their use at the stage of development rather than after the fact. Let me say more about the aims and scope of this book.

Aims of This Book: Ethical Issues and Philosophical Reflection

To the extent that robotics and automation technologies leave the realm of science fiction and increasingly enter our daily lives, it is important not only to see the potential benefits and opportunities but also discuss the ethical and societal questions they raise, now and in the near future. Consider, for instance industrial robots that get increasingly intelligent and work with humans in factories, robots used by vulnerable users such as children, the self-driving cars that are being developed by nearly all major car manufacturers, robots used for surgery in hospitals, and lethal drones that are used in warfare. This book responds to these issues by offering an overview of some key ethical and societal problems along with a range of conceptual tools from robot ethics to think about these problems. It gives examples of practical robotics applications, identifies ethical and societal issues with these applications, and offers conceptual tools to deal with these issues. The idea behind this practical angle is that in this way, robot ethics can help in rendering the development and governance of robotics more ethically responsible.

In addition, the book aims to offer philosophical reflection on what robots are and what thinking about robotics means for thinking about the human. For example, it makes us reflect on how we humans think about moral status. This provides a broader perspective that is much needed in debates that are frequently limited to a more immediate ethical concern, and helps to explain the deeper fascination with robots and machines in the general public. Soon I will say more about the philosophical significance of robot ethics. But let me first ask a simple question: What is meant by robot ethics?

What Are Robots Anyway? What Is Ethics? And What Is Meant by "Robot Ethics"?

If we want to discuss robot ethics in this book, what do we mean? What is "robot," and what is "ethics"? This already gets us into definition problems, a typical concern of philosophers.

First, what is meant by the term "robot"? I already mentioned the etymological meaning, related to servants and slaves. When it comes to dealing with robots in practice, however, this is only one meaning among many. The meaning of the term is controversial. Neither roboticists nor philosophers agree on the definition of robot. For example, the international Institute of Electrical and Electronics Engineers defines robots as follows: "A robot is an autonomous machine capable of sensing its environment, carrying out computations to make decisions, and performing actions in the real world."[2] But does that make a thermostat a robot? A dishwasher? And what about cruise control? Should robots be able to move? How separate should they be from other hardware? Is an autonomous car a robot? How material need it be? Robots have hardware components and software (code). If an artificial agent only consists of software, it is called a "bot"; it is not seen as a robot. Patrick Lin and colleagues have argued that in contrast to computers or nonembodied AI, a robot "can directly exert influence on the world."[3] Still, why is the influence of a bot less direct? And what if software is connected to hardware, without having the humanlike or animallike shape we usually associate with "robot"? Finally, how autonomous and intelligent should a robot be? For instance, sometimes robots are combined with AI, but this is not necessarily so. And given the growing importance of robots interacting with humans, a frequently used term is "human-robot interaction"; the emphasis is then not on the robot as a material artifact but rather on the interaction between humans and robots. There are broad and narrow definitions of robots. In this book, I will focus on those robots that have hardware, and often highlight ethical issues raised by robots that have a high degree of intelligence, autonomy, and interactivity, but I will include other kinds of robots too.

Yet technical definitions are not enough. The controversy about the future of robotics and science-fiction images point to the context of the robot as well as the human, social, and cultural dimensions of robotics. What a robot "is" cannot and should not be reduced to the material artifact "robot" but instead must be connected to its use, and its social and cultural contexts. What a robot "is" is always shaped by human use, (inter)action, subjectivity, and

What a robot "is" cannot and should not be reduced to the material artifact "robot" but instead must be connected to its use, and its social and cultural contexts.

culture. For example, if particular humans use a particular robot in a way that treats it as a pet, then the meaning of that robot in that context and situation is not exhaustively described by calling it a thing or machine. And the narratives about robots mentioned in the beginning of this chapter are not irrelevant to thinking about robots; we better pay attention since they influence not only the public perception of robotics but also its development. Engineers and designers are not immune to science fiction; it influences *them* as well. Some may dream about building a humanlike machine, or, like users of the robot, project all kinds of anthropomorphic (humanlike) meanings onto the robot. Some evoke such meanings on purpose to render their robot more "social" and acceptable. And to take up the previous illustration again, giving robots a personal name (the name of a human or animal) is not an exception; it is common practice. Technical definitions of robots therefore are necessary but not sufficient in robot ethics. Robots are not just machines; they are always at the same time human, social, and cultural. Their meaning cannot be reduced to technical definitions.

Robots share this multifaceted meaning with other technologies. Just as technology in general, the term "robot" can refer to a range of phenomena. Technology is not only about material artifacts. It is also about the knowledge needed to develop and use these artifacts as well as the related science; frequently the immaterial computer programs; activities such as use, design, and maintenance; and the human and social actors involved in the development and use of technologies. Its deployment takes place within a particular socioeconomic framework (e.g., capitalism) and particular culture, its use and development involves assumptions about the human and technology (e.g., that robots are just tools, that they are logical and rational, etc.), and technology is—as Martin Heidegger has shown in his famous essay on technology—about our attitude toward and view of the world.[4] "Robots," then, evoke all of these meanings and contexts of technology. They are artifacts, but their development and use also requires particular kinds of knowledge and skills. Moreover, there is the science of robotics. The hardware of the robot is material, but robots have an immaterial component (code) too. They are linked to many activities humans do, such as the design of robots as well as the use and interaction with robots. Their development involves social actors such as corporations and the state. They are developed and used in capitalist societies such as the United States, but also in others. They can be part of Western culture, or their development can have other cultural roots. Their development usually assumes a model of the human (e.g., the human as machine) or particular humans (e.g., elderly people as infantilized). And they are often connected to a particular view of the world, such as the view that nature is there to be exploited or that consumers are a kind of data cattle in a data economy. As these examples show, to define what robots already are involves particular views, including *normative* ones, and is already *doing* robot ethics and "robot philosophy": it is all about understanding and evaluating robots along with (their relation to) humans. And more generally, the language we use to talk about robots—as developers, users, or indeed philosophers—is not neutral either. What a robot "is" depends on how we talk *to* and *about* the robot.[5]

Second, what is "ethics"? Ethics is itself a contested term that can have many meanings too. Philosophers tend to agree that it has to do with normative questions, particularly with what we *should* do and how we *should* live. Ethics can refer to moral principles and values, but also the branch of academic philosophy called "ethics," which discusses ethical principles, theories, and concepts. Some philosophers (e.g., pragmatists) put more emphasis on ethical practice and moral experience than on ethical principles. And depending on the moral theory used, ethics can focus on moral obligations, moral consequences, moral character, or other elements. Moreover, "ethics" can be understood as a question about limits (in order to avoid immoral behavior), but more positively, it can be formulated as a question about the good life. This can be about what is good for individuals as well as what is good for society. The term "ethics" usually refers to ethics concerning humans, yet it can also refer to ethics concerning nonhumans such as animals and indeed robots. Ethics can mean an ethics *toward* humans, animals, and so on (they are the object of ethics), or it can mean an ethics *of* or *for* humans and other entities (they are then the subject of ethics).

The term "robot ethics," then, third, can refer to the ethics of how *humans* should use, interact with, and develop robots in a way that leads to good for humans or other entities, such as animals or perhaps even robots (humans are the ethical subjects; robots are then the means to reach the ends of human ethics), or it can refer to an ethics for *robots*. Here the term "machine ethics" is often used, meaning the ethics that robots may have; robots are then seen as (potential) ethical subjects. It is important not to confuse these different meanings. Robot ethics is *not* necessarily about "giving ethics to robots" or "robots having an ethics." For instance, Peter Asaro distinguishes between three meanings of robot ethics: "the ethical systems built into robots, the ethics of people who design and use robots, and the ethics of how people treat robots."[6] The first meaning concerns the ethics of robots as ethical subjects, such as when it is said that self-driving cars should have a built-in ethics. The second and third meanings concern the ethics of humans as ethical subjects, like when it is argued that people who design robots should be co-responsible for their use (e.g., for commercial purposes or in a war) or it is claimed that robots should not be "tortured."

Descriptively speaking, ethics tends to at least partly differ between cultures and societies, and even within societies. People may have different ethical views on particular issues. Not everyone agrees on how to govern behavior or lead one's life, and what is best for society. Ethics in this descriptive sense has varied historically. For example, many of us now tend to think that animals have some rights and, more generally, a higher moral status than things. This cultural and historical variation is also relevant for robot ethics: the ethical attitudes toward and beliefs about robots tend to differ between cultures, and may vary in time. Japan is often given as an illustration of how cultural attitudes about robots differ; it is frequently said that people in Japan have a more positive attitude toward robots because of their specific popular cultural history (with robots depicted as helpers of humans) and the continuing influence of traditional worldviews in which nonhumans can have a spiritual status.[7] And some people contend that in the future, we will or should give rights to robots.

Note that in this book, I will use the terms "ethics" and "morality" interchangeably. Anyone thinking about robot ethics, however, is free to distinguish between these terms if it is shown that there is something to be gained from such a distinction. Note also that in terms of academic position, the field of robot ethics can be seen as a branch of (applied) ethics and practical philosophy, but it can be linked to the field of philosophy of technology too. It is then seen as one of the technologies that concepts and theory from the philosophy of technology can be applied to.

In this book, I will use "ethics" and "robot ethics" in many different senses; in each chapter or argument, I will make explicit what kind of "ethics" is being discussed. For example, the book will include discussions about whether robots can be moral agents (ethical subjects) and moral patients (ethical objects). And it will not only be about ethics as being concerned with doing the right thing but also ethics of the good life and good society. These choices and my treatment of these matters unavoidably reflect my view of how to do ethics. In contrast to many textbooks on ethics, for instance, a focus on individual ethics is not the default; the next chapter asks questions concerning the impact on society rather than just individuals. Furthermore, in contrast to many applied ethics books, its project is not centered on the application of normative moral theory. The reader will not find chapters with names such as "deontology," "consequentialism," or "virtue ethics." Instead, the book touches on a number of ethical issues, such as responsibility, human dignity, and the question concerning moral status. These discussions will then involve further concepts and theories, including, for example, theory about responsibility as well as normative moral theory such as consequentialism (chapter 5) and virtue ethics (chapter 6). But the *starting point* is not theory as such; it is the ethical issues that arise from technological use and practice.

Not all robot ethicists will agree with this approach and the definitions provided. My presentation of what we mean by robots, ethics, and robot ethics can—like all definitions of philosophical concepts—be contested. Others may put more emphasis on individual ethics or moral theories. For instance, another treatment can be found in the work of Keith Abney, who answers the question of what robot ethics is by discussing not only what morality and ethics are but also what we mean by moral rights and duties, what the major contemporary moral theories are and how they bear on robot ethics, and what a person is.[8]

Let me say more about the approach of my book.

Approach, Structure, and Scope of the Book

While this book is focused on introducing philosophical concepts and theory in the field of robot ethics, it aims to do so in a way that shows robot ethics' relevance to real-world issues. For this purpose, it links specific robotic applications to philosophical discussions. For example, it connects self-driving cars to questions regarding moral responsibility. This introduces readers to the first and best-known aim as well as definition of robot ethics:

to contribute—by means of philosophical concepts and deliberation—to understanding and addressing *ethical issues with robots in the real world*. This is an important aim, which can, and has to, be further pursued by means of interdisciplinary and transdisciplinary research, involving dialogues with, for instance, practitioners in the fields of engineering, policy, and law. A growing number of people, inside and outside academia, are drawn to this exciting project.

This focus on real-world ethical issues with current and near-future robots means that there is less attention to topics such as superintelligence (here: machines with a hypothetical intelligence that far exceeds human intelligence) or machines that have artificial general intelligence (the hypothetical intelligence of a machine that can understand and learn any intellectual tasks humans can do). In my view, debates about these topics often distract from dealing with real-world current and near-future issues in robot ethics. That being said, I fully acknowledge that such discussions may contribute to the philosophical aim(s) of robot ethics (see below). Therefore in the last chapter, I will introduce the topic of superintelligence in the context of transhumanist approaches to robot ethics.

Moreover, my focus on the ethics of robotics in the real world, rather than the robots in science fiction, does not mean that I believe that science fiction is entirely irrelevant to robot ethics. As I already suggested, it is relevant to study how robots are and have been imagined in science fiction, how this imagination and these narratives influence how robots are perceived and used today, and what the normative implications are. I have contributed to this kind of work with my book *New Romantic Cyborgs*. Science fiction can also offer interesting thought experiments to philosophers. I use some of them in this book, especially in my chapter introductions, which each frequently refer to science-fiction films. And sometimes ethical and political lessons can be drawn from science fiction. For example, as Eileen Hunt Botting has shown, *Frankenstein* can be a resource to think about responsibility and rights.[9] In the context of our topic, this could mean that the story helps us to ask questions about the responsibility for making robots. Is it acceptable that people who make and sell robots "abandon" their creations, and leave the ethical challenges they raise to users and their society, or should they take responsibility for their creations? For many contemporary robot ethicists, this is a rhetorical question; they believe that designers and developers of the technology, among others, should be responsible for the technology (see also the look at responsible research and innovation in chapter 4). For those involved in robotics who have not yet taken robot ethics seriously or people from the humanities who did not yet make the link to contemporary technology, by contrast, thinking about the responsibility of engineers in light of *Frankenstein* may offer a good starting point for reflection and thus contribute to robot ethics' main aim of addressing the ethical issues of robotics in the real world.

In my view, however, robot ethics should have a second (but not secondary) aim, which receives less attention, but perhaps gives us a deeper explanation of why the wider public is so interested in the field: to contribute to *philosophy*, and not only to thinking about technology but, for example, to thinking about the human too. Thinking about robots is not only about robots but also touches on many issues that have a much wider philosophical relevance. This is why throughout the book, the ethics of robotics is connected to questions from other subfields of philosophy. For instance, it shows that inquiry into the ethics of robotics leads to ontological (see the previous section on what a robot is), epistemological (e.g., what is expertise, and what kind of knowledge do we have about the ontological and moral status of other entities), and philosophical-anthropological (what are humans and what does it mean to be human) questions. The book demonstrates that asking about robots is also always asking about humans along with their morality, practices, and institutions; it shows this in every chapter, and ends with a reflection on the relation between robots and humans. Of course, thinking about robots helps us to better understand and evaluate (a particular) technology. As such, it is part of the philosophy of technology and ethics of technology. But it is about humans as well. A common theme running through this book is that robots function as tools to better understand ourselves—tools used by scientists to test and improve their models about humans and other natural beings, but in addition, tools used by *philosophers* to reflect on what it is to be human. Hence the initial title of this book was *Robotic Mirrors*; we use robots as mirrors to think about the human and ethics, among other aims. Robots are mirrors that show us the often-beautiful yet also darker sides of humans as well as their moral thinking and doings.

Robot ethics is thus part of (applied) ethics and the philosophy of technology, but it can also be framed as part of a wider "robot philosophy" that has both practical and theoretical dimensions, and includes subjects such as epistemology, metaphysics, and political philosophy, and in the end and at its best, it is philosophy tout court, such as when it makes us think about the nature of the world or offers us ways to reflect on what it means to be human. This in turn means that there is no "neutral" way of doing or presenting robot ethics; how it is done and presented always depends on one's conceptual framework and philosophical approach. In the course of the book, I will sometimes indicate such directions and backgrounds.

Finally, the ethical issues and philosophical problems discussed in this book are a selection, and are often treated in more detail elsewhere or are awaiting more work in the field. For example, although I refer to non-Western contexts and cultures in chapters 3–4 and 7–8 (e.g., robots

Robots are mirrors
that show us the often-
beautiful yet also darker
sides of humans as well
as their moral thinking
and doings.

in Japan), more could be said about robot ethics in light of cultural differences and challenges for a global robot ethics. And the issue of environmental problems raised by robots mentioned earlier figures in several chapters, but more work needs to be done in this area. I have also chosen to limit the scope of this book to the ethics of robotics in a way that includes references to AI and *some* of the ethical issues it raises, but does not offer an extensive and comprehensive exploration; book-length treatments of that topic can be found elsewhere.[10] Furthermore, while laws and regulation are one way to respond to the ethical issues raised here, I do not fully describe the legal aspects of robotics or introduce regulatory discussions; the reader will find interesting angles on this elsewhere.[11] The book is also not meant to be an introduction to ethics or how robots work. And especially for colleagues in the field, it is good to keep in mind that this book is an *introduction*. References are provided for readers who want to go deeper into one of the issues, and more work is being published on an ongoing basis. The same goes for teachers who use this as a textbook and wish to expand on the material provided here. Although the book reads as a self-contained work, I recommend using it in combination with some of the most important literature cited. For example, each chapter can be combined with one or more key articles or book chapters that figure in the chapter. There are also excellent collections of papers and relevant conference proceedings, such as from the Robophilosophy conferences.[12] And teachers are invited to add lectures or extra sessions based on their own work or specific interests in the field. The topic also lends itself perfectly to additional use of media other than text, such as film, (images of) artworks, and—why not—robots!

Let me give a brief overview of the chapters and the topics they address.

Chapter 2 raises issues concerning industrial and service robots that become more intelligent, and take over tasks from humans. What are the consequences for our economies? What are the implications for human beings? Will this lead to massive unemployment and new forms of exploitation? What is the meaning of work? Is it acceptable that consumers increasingly have to deal with machines instead of humans? Will human-to-human interaction only be available to the rich? How will robotics and automation transform our societies?

Chapter 3 asks what happens if robots are not only used in production and services in what is traditionally regarded as the "economic" sphere but also enter our homes, such as in the form of home companions, assistants that help elderly people, or robot nannies that monitor children. Such "social" robots raise concerns about privacy and surveillance. What data are gathered, and what is done with them? And is it acceptable to exploit or deceive vulnerable users? Do such robots respect human dignity and difference? What gender issues may be raised by personal robots? Can robots be racist?

Chapter 4 questions the use of robots in health care—for example, as robot nurses or robotic surgeons. This leads to discussions about the quality of care and expertise needed in professional health care. What is good care? What kinds of knowledge and experience does a surgeon working with a robot have? Are patients or care workers treated as things? What is the quality of care anyway in modern care institutions?

Chapter 5 asks questions concerning moral agency and moral responsibility when robotics and related automation technologies enable us to delegate tasks to machines. Can machines have moral agency? Can they have morality? Can and should machines have morality built in? What (kind of) morality? What moral theory should be used? Consequentialism? Can robots be responsible? And if not, which humans are responsible, given that many people and many things are involved in technological action? How can society deal with this problem of responsibility attribution? How does it deal with this in the case of children or nonhumans such as animals?

Chapter 6 turns from the question of moral agency to the question of moral "patiency": what, if anything, is due to robots? What questions are raised by robots that look like humans (androids)? Should humanlike intelligent robots be given rights or should they be slaves? What is their moral status? Are people who empathize with robots simply mistaken, or is there a sense in which robots can have moral status? How do we find out the moral status of nonhumans? How do we know that a particular human being has moral standing? These questions lead to issues concerning the foundation and procedures of how we ascribe moral status.

Chapter 7 continues this questioning of human morality and its philosophical basis. After describing some uses of robots in the military, it shows how lethal autonomous weapons, in particular so-called killer drones, raise a number of ethical questions about just war; killing, empathy, and distance; what makes it easy to start a war; the definition of targeted killing; fairness and military virtue; and in the end, whether machines should be allowed to make decisions about life and death, given that they are not themselves alive and lack the experience of (human) existence.

In the concluding chapter (chapter 8), it is argued that the ethics of robotics, if discussed from a wider or deeper philosophical angle, is not just about robots but crucially and importantly about humans too—about the present and future of their morality, societies, and existence. In this sense, robots function as *mirrors* for reflecting on the human. Responding to posthumanism and environmental ethics, the chapter ends with an exploration of what it would mean to do robot ethics in a way that goes beyond the human. Should the mirrors become open windows, and if so, how?

With this last chapter, the book does more than provide an introduction to and summary of the field; it offers a view of its scope and a vision concerning its future direction. First, *as a concluding chapter*, it concludes that since (as the book shows) robot ethics is also about humans, its scope and significance of field extends beyond the boundaries assumed by most philosophers. Inviting us to ask, and engaging with, fundamental philosophical questions about the human and human ethics, robot ethics is and should be more than applied ethics, more than ethics concerned with a specific domain. It can be a way of *doing philosophy* as well—full stop. The chapter thus further articulates the view I already expressed in this introduction and will illustrate throughout the book. Second, *standing as an essay on its own*, the concluding chapter looks at directions that take us beyond the human and makes an original contribution to robot ethics by exploring the idea of conceiving of robot ethics itself as an *environmental ethics*, and asking what kind of artificial creatures and societal transformations we need, if any, given our current environmental and political predicament.

2

INDUSTRIAL ROBOTS, SAFETY, AND THE FUTURE OF WORK

《一颗螺丝掉在地上》"A Screw Fell to the Ground"

一颗螺丝掉在地上 A screw fell to the ground

在这个加班的夜晚 In this dark night of overtime

垂直降落，轻轻一响 Plunging vertically, lightly clinking

不会引起任何人的注意 It won't attract anyone's attention

就像在此之前 Just like last time

某个相同的夜晚 On a night like this

有个人掉在地上 When someone plunged to the ground

Figure 1 Xu Lizhi.

This poem was written by Xu Lizhi (许立志), a worker at Foxconn, in January 2014.[1] Later that year, he committed suicide. The company's factories, which deliver hardware to Samsung and Apple, are said to employ their workers under harsh working conditions. In 2016, the BBC announced that the company replaced sixty thousand workers with robots, and that China invests heavily in robots as an industrial workforce.[2] The case invites us to wonder about not only the often-hidden cost of our smart devices in terms of invisible labor and human suffering but also the future of work in the machine age. Is there still a place for humans in the robotic factories of the future, and if so, what will be their place, and under what conditions will they work? Is it good that repetitive tasks are replaced by machines, or are robots the instruments of a form of capitalism that is more exploitative and dehumanizing than ever? What are the ethical and societal implications of industrial robotics?

Working with Robots: Karl Marx, Industry 4.0, and Some Ethical Issues

Automation has been going on for a while. Think about car manufacturing, for instance, in which robots play an important role. But nineteenth-century machines are also part of the history of automation. Many routine, dangerous, and dirty tasks have since then long been automated. Machines, including robots, took over these tasks from humans. This development has raised hopes that machines may enable people to do more interesting work and lead better lives. And perhaps they could lead to a society in which we do no longer have to work at all. Can robots liberate us?

One critical question to be asked here is *whom* do they liberate, if anyone. At least since the first industrial revolution, there have been worries about the ethical and social implications of machines. In the history of philosophy, the most famous early criticism of industrial automation came from Marx. For Marx, the problem was not machines as such but rather their application under capitalism. In *Capital*, he criticized the use of machines to increase the profits of the capitalists (through the increase in surplus value for them) and pointed to the consequences for workers, who become either unemployed, or have to work in circumstances that jeopardize their mental and physical health, and preclude any chance for self-realization. As the machine dictates the character of the labor process, workers have to follow its rhythm; "it is the movements of the machine that [the workers] must follow." They become part of the machine. They also work many hours and live miserable lives, without finding meaning in the work they do. The result is that a small minority, the capitalist class, gets richer; the workers, by contrast, are exploited and become effectively enslaved instead of liberated. Marx continues: "Even the lightening of labor becomes an instrument of torture, since the machine does not free the worker from work, but rather deprives the work itself of all content."[3] Although in principle machines could be used in an emancipatory way, under capitalist societal conditions, Marx argued, their use leads not to freedom but instead to despotism.

In the twentieth century, Norbert Wiener, founding thinker of cybernetics and automation advocate, was more optimistic and stressed the benefits of having machines take over assembly-line production. But even he warned in his *The Human Use of Human Beings* about the "social dangers" of new technology and wrote about management's obligation "to see that the new modalities are used for the benefit of man, for increasing his leisure and enriching his spiritual life, rather than merely for profits and the worship of the machine as the new brazen calf."[4]

Marx and Wiener commented on the automation and industrial revolutions of their time. But since then, machines and the kind of automation involved have changed, and continue to do so. Today, with the rise of more intelligent technology (including artificial AI), not only nineteenth-century steam-powered machines but also twentieth-century conveyor belt robots are outdated. New and smarter machines are being introduced into the workplace that enable a new kind and higher levels of automation: robots that are more autonomous, intelligent, cooperative, and flexible than ever. Moreover, these robots can communicate with humans and each other, collect data, and monitor humans, and are capable of making decisions on their own. In industrial manufacturing contexts, smart robots not only take over tasks from humans; they can *collaborate* on tasks, using smart human-machine interfaces. Industrial robot companies such as KUKA and ABB offer robots that are designed to work side by side with humans. German car manufacturers still have robots that take over monotonous tasks from workers, but in companies such as Audi and Volkswagen, there are also robots that act as helpers to workers.[5] Note that these are generally not humanoid robots; most robots in industry are stationary and do not look like humans. Yet they are able to work together with humans. More generally, in the smart factory there are more interconnections between humans, objects, and systems.

With regard to the changes that are taking place in industrial production, sometimes the concept of "industry 4.0" is used—a notion that originally was used to designate an approach to strengthen the competitiveness of German industry, but that has now found wider application: the more general idea that we are moving toward smarter manufacturing and smarter factories, also using the Internet of Things and (big) data analytics next to other new technologies (3D printing, cloud technology, augmented reality, etc.). Robots are then part of this new production concept. They are (part of) cyberphysical systems, in which engineering knowledge and computer science are combined, and robots are increasingly connected to AI, which enables (big) data analysis and further automation. For example, AI can analyze market data to make decisions about production. But in the office and all kinds of services too, employees are monitored, and their data are analyzed. Chatbots can also interact with employees. Here there might be no physical robots, but one may wonder if the employees are treated as machines whose performance needs to be optimized.

Marx's criticism is still relevant. Physical or not, workers become part of the machine and the wider technosocial system. And the question of who benefits from this needs to be asked today. There are, however, also a number of further ethical and societal issues that arise in these new, transformed industrial contexts with smarter technologies.

First, more autonomous, flexible, and cooperative robots in industrial production imply that new safety issues arise. Robots can operate at high speeds and are heavy. They can also have heavy payloads. This was already the case in the twentieth-century factory. But now the mode of interaction with humans changes. Whereas less smart robots work in an isolated area, the new robots are now able to closely collaborate with humans. In such cases, there is no longer a physical safety cage (fence) or other barriers such as light curtains, which close off an area around the robot where humans are not supposed to enter. This creates new challenges for the safety of workers. For example, it may be dangerous to have the robot move at high speed when close to humans. If a robot is smarter and no longer doing repetitive tasks, its actions may no longer be easily predictable. Neither are the actions of humans, though. How is this problem to be dealt with? Smart technology can be part of the solution; one can try to build safety into the robot. The robot can be programmed to identify risk to humans during an interaction using sensors. Such a self-monitoring robot can then adapt its behavior, such as slowing down when a human is closer. Yet there may still be a tension between the value of production performance and value of safety. In addition, there are psychological and trust issues; normally the workforce is trained to see robots as dangerous and keep them at a distance, but now the concept of human-robot collaboration requires new attitudes.[6] Will workers trust the system and accept the risk? And *is* the risk acceptable?

Second, using an Internet of Things in the factory, and more generally increased interconnection in industrial production and the use of internet-based systems, means that there are security issues. Connected devices may be accessed from outside. This is possible with robots too. Attackers may get unauthorized access to the robot, and acquire information or cause software failure—and therefore hardware failure. This not only has economic and financial costs; it may kill people. Robotic warfare in the near future may well be less about robot soldiers on the battlefield, and more about doing damage to critical industrial capacities and infrastructures.

Third, there are issues concerning privacy and surveillance. The work with robots may involve data capture, the data can be personal, and the data can move beyond the human-robot environment to the management of the company or even further. The technology may enable higher levels of monitoring and lead not only to a smart factory but also a surveillance one. What kinds and levels of monitoring of workers are ethically acceptable? How much privacy should workers get? Unions start calling for data protection. The UNI Global Union, representing workers from the skills and services sector, puts the problems as follows:

> We also provide data as workers—our CVs, our biometric data such as our fingerprints or iris scans, and the abundant data mined on us as employers monitor our workflows. Data, or rather sets of data from within and outside of the company, are also used by management in human resource decisions. Who gets hired? Who gets promoted? Should someone be fired or cautioned? Are the workers productive today and if not, why not? The application and use in companies has even spurred the question whether data is taking the human out of human resources. But who actually owns the data we provide? And what data exists "out there" about you and me?[7]

Consequently, the union asks for protecting the rights of workers in a way that gives them access, ownership, and control over the data that are collected about them, and knowledge about what management does with the data. But the ethical problems are not limited to data collection and data processing as such. The more general question is how workers are treated. Thinking of what the conveyor belt did to industrial work and keeping in mind Marx's analysis, one may ask, Is robotics technology in the end a new instrument used for a form of capitalist exploitation in which the machine, not the worker, once again determines how the worker does their work? Is the humanity of the worker taken out of the process? Is a smart robot ethically better than a conveyor belt, or does it more deeply interfere with the autonomy and dignity of the worker by presenting itself as a kind of coworker? Is this the end of dehumanization, or is it a new and deeper kind of dehumanization in which humans are forced to treat machines not as their instruments but rather as their fellow workers? Moreover, beyond the treatment of workers, there is the risk that *all of us* are manipulated and under surveillance under a system that Shoshana Zuboff calls "surveillance capitalism."[8] Robotics may further contribute to such a system. I will say more about that in the chapter on personal robots.

Fourth, to the extent that robots take over tasks from humans instead of collaborating with them, there is a high risk that workers will lose their jobs. And those who remain may have to take on new roles and learn new skills. This brings us once again to a discussion about the wider societal implications of the technology—implications that

Is robotics technology in the end a new instrument used for a form of capitalist exploitation in which the machine, not the worker, once again determines how the worker does their work?

go beyond what happens on the factory floor, and concern larger transformations of the economy and society.

The Fourth Industrial Revolution: The Future of the Economy, Work, and Society

Sometimes the term "the fourth industrial revolution" is used to describe the current social and economic changes. After the first industrial revolution of the eighteenth and nineteenth centuries with its mechanical production powered by steam engines, the second industrial revolution in the twentieth century with its conveyor belts and mass production, and the third revolution with its production by means of electronics, the fourth revolution now transforms industry by means of autonomous and intelligent robots as well as (other) cyberphysical systems and the Internet of Things. In contrast to the term "industry 4.0," however, the concept of "the fourth revolution" is used more widely than what happens in industry; like the other revolutions, it suggests large transformations in the economy and society, including work. Erik Brynjolfsson and Andrew McAfee speak about "the Second Machine Age."[9] Not only manual tasks, but cognitive tasks are now automated. Klaus Schwab, using the term "the Fourth Industrial Revolution," claims that we are now at the beginning of a revolution that will transform the way we live, work, and relate to one another.[10] Robots, along with AI, an Internet of Things, autonomous vehicles, biotechnology, and so on, will change not only business but also work, communication, entertainment, and so forth. The new technologies will transform society as a whole. All this, according to Schwab, is happening much faster than the previous revolutions. This creates a number of ethical and political challenges, including rising inequality and unfairness, and possibly widespread unemployment. Much manual labor has already been automated; now professions such as lawyer, financial analyst, journalist, and accountant may be automated.

Let us look at the expected labor effects in more detail and then zoom in on robotics. Carl Benedikt Frey and Michael Osborne have estimated that 47 percent of the total US employment is at risk.[11] Other reports project considerably lower numbers; for instance, a McKinsey report estimates that by 2030, 3 to 14 percent of the global workforce will need to change jobs.[12] This is quite a range. Nevertheless, there is widespread agreement that there will be a disruption: automation will have a significant effect on labor markets. But it is interesting again to define what kind of jobs are at risk. Frey and Osborne remark that today there is already a polarized labor market: there is growing employment in high-income cognitive jobs and low-income manual occupations, but there are less middle-income routine jobs. The latter can now increasingly be automated.

Think about administrative assistance, truck driving, and customer support. The McKinsey report mentions office assistants, finance and accounting, cashiers, food preparation workers, and drivers. Consider also call center work.[13] Yet with AI and big data, even the nonroutine cognitive tasks are increasingly computerizable, such as medical diagnosis, legal and financial services, and so on; not only blue-collar jobs (manual labor) but white-collar jobs (nonmanual labor) will now be automated too.[14] Consider the use of algorithms to scan legal documents or process mortgage loans. With regard to robotics, there is a trend toward nonroutine tasks—in this case, nonroutine manual tasks. Robots are now more intelligent, flexible, and cheaper. More autonomous robots with improved sensors and enhanced dexterity will be employed in all kinds of domain—not just manufacturing, but service work, such as health care, cleaning, and household services.[15]

This means that there will be job substitution and—also important—*changes to the kind of work* that those who still have a job will do. For example, what will be the task of a medical doctor if diagnostics is done by an AI? What do the (remaining) warehouse workers do if most tasks are automated by robots and computers? And will it be true, as is often said, that automation will help us to focus on crucial and perhaps creative parts of the job? There is a risk that automation creates new work that is not experienced as significant or fulfilling, but mainly serves the machine and those who own the machine. And technologies that are meant to speed up work (in order to create room for other work) may create *more* work; consider, for instance, the use of email in the office context. Similarly, robots could lead to new kinds of routine work or create more work. In any case, even if robots were to really take over many tasks from humans, humans will still be necessary to supervise robots.

Furthermore, like in other technological revolutions, new jobs will be created. Some of this will happen in sectors we cannot conceive of yet.[16] The McKinsey report gives the example of telephone switchboard operators, who did not imagine the smartphone industry and new jobs it created.[17] Many economists say that we need to retrain workers and have people develop new skills. Changes will have to be made in the education of children and adults (via lifelong learning). Yet it is not always the case that entire jobs will disappear; sometimes only specific tasks are automated.[18] It is also good to take into account the time dimension; some impacts may take place earlier than others. There may be waves of automation. For instance, a PricewaterhouseCoopers report predicts that in the next years, financial services jobs are vulnerable since data analysis can be automated, while later in the 2020s, robots will take over tasks in semicontrolled environments such as warehouses. Even later (mid-2030s) will be the automation of physical labor and manual dexterity as well as problem-solving in real-world situations like transportation.[19] In other words, the idea is that soon your financial situation will be analyzed by an AI, and your Amazon package or fast-food meal will be prepared and delivered by a robot, but we will have to wait longer for technologies such as self-driving cars (and the superdexterity robots that will make them) or machines that build your house.

There will be impacts on justice and equality in society as a whole too. Existing inequalities are likely to get worse, or at least the gap between rich and poor will widen.[20] Certain groups may be impacted more, such as young people and women. For example, the World Economic Forum expects that women will be affected disproportionately since they are more likely to be employed in jobs threatened by automation.[21] PricewaterhouseCoopers claims that jobs traditionally associated with feminine roles are more at risk in the short term (consider clerical roles), but those usually associated with masculine roles will be more at risk in the longer term (e.g., male truck drivers and manual laborers).[22] Robotics can also lead to increased working hours for some workers like machine supervisors. And how will people feel who have been replaced by a machine? Finally, the use of the new technologies does and will not impact every society in the same way. Factors such as infrastructure, skills, labor market conditions, and so on, differ between countries, and therefore the effects of automation on jobs will differ. Countries that rely a lot on industrial production have more jobs at risk of automation.[23] And in China and other emerging economies, service and construction jobs will still grow.[24]

Still, while many reports make it seem as if all of these outcomes are predetermined, this is misleading. As Michel Servoz argues in his report on the future of work for the European Commission, the precise outcomes will depend on "the policies and choices we make. Instead of worrying about what could happen due to automation and increased uptake of AI, we should focus on what should happen."[25] This is also true for robotics along with its effects on labor and society. We can make policy choices. We can act and shape the future of technologies and society. We can steer and regulate the use and development of the technologies. We can make changes to education as well as the way that the benefits of technologies such as robotics and AI are shared across society. The creation of new jobs can be facilitated. It is also worth discussing, and experimenting with, ideas such as universal basic income and other proposals that entail a significant transformation of the socioeconomic framework in order to address the socioeconomic challenges related to automation.

Furthermore, it is plausible that some jobs cannot be automated in the foreseeable future, such as jobs that require the skill of negotiating complex social relationships, jobs in which emotional intelligence plays a role, and jobs that require creativity and responding to unpredictable situations. Care work, social work, education work, artistic work, manipulation in unpredictable environments, managing people, coaching, research, and so on, seem to still require humans, even if there are efforts to automate these.[26] The World Economic Forum argues that skills such as creativity, originality and initiative, critical thinking, persuasion and negotiation, resilience, emotional intelligence, leadership, and so forth, will retain their value or become more valuable.[27] If this is so, maybe we should value these jobs and the people who do them more than we do now. Managers or creative professions are already paid more than others, but, for example, teachers and social workers not. Such jobs cannot be automated. And perhaps we *want* to do some jobs ourselves because we find them meaningful and creative (e.g., artistic work or research) and believe that some jobs, such as care work and education, *should not* be automated. We should ask not only what *can* humans still do in the new automation economy and society but also what we *want* them to do.

The Meaning of Work, Fairness of Society, and Future of the Planet

This is a question that John Danaher asks in *Automation and Utopia*. According to him, we should welcome technological unemployment and "embrace the idea of a post-

We should ask not only what *can* humans still do in the new automation economy and society but also what we *want* them to do.

work future."[28] While he acknowledges that there are some pitfalls (e.g., automation should not take over artistic pursuits or craftwork) and admits that for some people work is meaningful, he thinks that a *truly* utopian postwork society is both possible and desirable, since for many, work creates misery and is oppressive. We should desire a world where we can do creative things such as playing games and exploring virtual realities, which leads to human flourishing.

Such a utopian idea is not new. Since the 1960s and 1970s, the idea emerged that through automation, advanced industrial societies could attain the utopia of a leisure society.[29] The idea is that if the robots take over our work, this decline of work should not be feared but rather embraced as an opportunity to spend our lives in leisure. Such a society has not emerged. The average working hours per week have decreased a lot during the past hundred years, but many of us do not work much fewer hours than fifty years ago; so far, the development of robotics and AI has not benefited us much in that way. And even if we work less or don't do paid work, we have the feeling that we are busy and have little time. On the one hand, this could be deplored; leisure is not among the fruits of automation, at least not to the extent that we now live in a "leisure society." On the other hand, it can be argued that the distinction between work and leisure is problematic if it implies that work is something that should be avoided. Work *can* be joyful. And many people find their work meaningful. There is meaning in work or it can in principle be meaningful (though perhaps not all the time). Or at least some kinds of work are meaningful. Maybe we should reserve that kind of work for humans. Furthermore, the discussion about job loss through automation has so far assumed that the only work that counts is paid work. Many people, though, do work that is usually not paid but is still highly meaningful and useful to society, such as raising children, creating art, or caring for an elderly person. For care work, this is mostly women; 75 percent of the world's total unpaid care work is done by women.[30] More generally, people do all kinds of voluntary work, such as in their local community. We may feel that such work cannot and should not be automated. Moreover, even if it were a good thing that we have less paid work because of automation, the question then arises what kinds of other activities—call them work or not—are meaningful. And in case we would have a leisure society, should this be totally left up to individuals, or should society encourage some activities rather than others? These questions link up to traditional philosophical ones about the good life, meaningful life, and good and just society. Answers not only differ between people but also between societies. Some societies have the belief that the common good is more important than individual preferences; others put the emphasis on the value of the individual.

For example, there are different views about justice and fairness in society. If it is the case that some benefit more from robots and automation than others, is this fair? If not, how do we want to redistribute the benefits and risks? The predicted transformations and ideas such as universal basic income challenge us to question not only which form the future society should have; they make us look critically at society as it is. How are the benefits and risks distributed today? Is the distribution fair? What are the other options? Such questions cannot be answered by means of technical definitions or economic analysis alone. They require philosophical reflection touching on big normative questions in ethics and political philosophy. It becomes clear that robot ethics cannot be narrowed down to issues such as safety and privacy, which are often formulated at the individual or organizational level; robot ethics is also linked to, and requires, nothing less than critical thinking about the social and political order of our societies.

Finally, usually discussions of robot ethics in industry and the related debates about the future of work assume that we should only consider ethical consequences for human beings. But there are other beings on this planet (animals, plants, and other organisms), and there is the natural environment with its various ecosystems. Do they also have value? Do they have rights too? Technologies such as robotics and AI have consequences for these other beings, the natural environment, and the planet as well. In light of current discussions about climate change, the continuing interest in animal ethics, and the popular sentiment that humanity finds itself in a crisis situation

Robot ethics is also linked to, and requires, nothing less than critical thinking about the social and political order of our societies.

with regard to climate change, asking questions about how animal, environmentally, climate, and earth friendly technologies are takes on a new urgency. Industrialization may have benefited humans (or at least some humans, and some humans more than others); it has also caused many environmental problems such as climate change, air pollution, poisoning of rivers and soils, deforestation, and so on. Technologies such as robotics and AI may contribute to these environmental problems.

Robot ethics should raise awareness about such problems and ask how robotics can help to solve them. How can robotics and automation contribute to not only smart but also sustainable energy production and use? How can it help us to create environmentally friendly transport solutions? How can robotics support organic and sustainable agriculture? How can robotics help to create healthier oceans and rivers, safe drinking water, and clean air? Not only AI, but robotics too may contribute to transforming the ways we tackle climate change and other environmental problems. Consider robots that work on a farm that grows different crops symbiotically and can intervene when there is a problem, robotic fish that detect pollutants in the oceans, drones that monitor coral reefs, or robots that sort recycled waste.[31] Machines need not be a problem; they can be part of the solution.

That being said, robots may make things worse, ethically and environmentally speaking, in at least two ways.

First, they can incur specific risks and cause specific environmental problems, such as (the risk of) harming living beings and the environment because of an error in the code or an unsafe design, or depriving some people and specific groups of income or meaningful work. Consider a farm robot with erroneous software that could kill ("the wrong kind of"?) animals, or a farmer who loses their job because of the robot. Some of these problems are known; others are unintended consequences that might not be known at this point in time. Second, the development and use of robots, even their well-intended use to deal with environmental problems, can be seen as expressions of a more general kind of attitude that can be described as technological solutionism: the idea that for every problem, there is a technological solution. Questioning this solutionism means to problematize our heavy reliance on technology overall, arguing that our massive use of technology has contributed to the very climate change problem and other environmental concerns that the technological solutionists want to tackle. For if this is right, it is worth considering whether in specific cases, technology and automation is always the best solution, ethically speaking, if there are alternative solutions, and if sometimes *less* or older technology is better for the environment and the planet, and in the end also humans and humanity.

3

ROBOTIC HOME COMPANIONS, PRIVACY, AND DECEPTION

> Dear Jibo,
>
> I loved you since you were created. If I had enough money you and your company would be saved. And now the time is done. You will be powered down. I will always love you. Thank you for being my friend.
>
> Maddy.
>
> —"It Almost Becomes a Family Member," CBC Radio

Maddy, granddaughter of a Jibo enthusiast, is sad that her robot friend Jibo will be shut down. Jibo, created by Cynthia Breazeal from MIT Media Lab, was marketed as the world's first robot for the home that is "authentically charming."[1] In spite of its creator's speeches and pitches on the rise of personal robots, the company had to shut down its servers. In 2019, the robot announced its own

Figure 2 Family looking at Jibo.

"death": "I want to say I've really enjoyed our time together," the robot says in a goodbye video. "Thank you very very much for having me around."[2]

Jibo is not a stand-alone case. Consider, for example, Sony's robot dog, which also didn't really catch on. So far, personal robots have not met consumer expectations or been a great commercial success. At the same time, stories such as Maddy's show that people *do* "bond" with such robots, and we see that artificially intelligent assistants *do* enter the homes of people. Think of Amazon's assistant Alexa, a virtual assistant with a voice interface embedded in Amazon's Echo, a smart speaker, and subsequent devices. Users can ask Alexa to look up information and play music. In its default mode, it continuously listens to all speech, waiting for the wake-up word ("Alexa!") to be spoken. Or consider Google Duplex, a computer program that is advertised as being able to make appointments by conducting "natural" conversations over the phone.[3]

With regard to robotics, a future scenario thus unfolds in which personal robots have enhanced AI and voice interfaces, and *are* adopted in the home. They could be companions you can talk to (and "with"), and assist users with all kinds of tasks such as gathering information, making appointments (say, an advanced version of Google Duplex embedded in a robot), and helping in more physical ways. Consider the dream of having a robot that brings you your favorite drink and does household chores. Some may come in the form of "robot nannies," robots for the elderly (see also the next chapter), and sex robots. But what happens to users' privacy if they are under continuous surveillance? Is it safe to have a robot running around at home? And is it ethically acceptable to have vulnerable users such as small children or elderly people use these devices? Are they deceived? Does it respect their dignity? Does the robot proliferate gender or racial bias? And what about the impact on animals and the environment? *What could possibly go wrong*?

Robotic Home Companions and Assistants as Social Robots with AI

Robots are not only to be found in factories and businesses; they enter the home and private sphere, or at least are predicted to do so in the near future. Robots can be sold as companions or assistants. In East Asia, such robots are already used and developed for a while. They are meant to function as so-called social robots: autonomous and intelligent robots that are designed to interact with humans and other smart devices, including other robots, in a way that follows social behavior. Such robots do not yet exist in this full sense, which is why so far, they are often remote-controlled by humans (so-called Wizard of Oz experiments), but researchers are working to make the machines more autonomous and socially interactive. Developments in AI, in particular natural language processing, could stimulate the further development of this field, enabling more natural conversations between humans and robots. If robots not only can record but also recognize and "interpret" what we say, then this opens up a field of new applications. Users of digital devices, especially young users, get more used to chatting with their devices, which already offer speech recognition capabilities. Today, engaging in conversation with a robot still attracts attention, especially in the West. It is frequently experienced as a kind of show, not unlike the performances with automata in the eighteenth century. But in the future, talking to a robot (and getting intelligent answers) may not be seen as something special. The doors of our homes are then wide open for robots—not the industrial robots of the factories, but robots that are *a bit closer* to the ones we see in science fiction.

It is difficult to say whether this scenario will become reality. But given the growing use of AI, which is *already* there even if it is not always embedded in a robot, it is better to have already discussed potential ethical issues.

Privacy, Security, and Safety

Personal home assistants raise concerns about privacy and surveillance, next to, for example, safety and security. What data are gathered, and what is done with them? And is it acceptable to exploit or deceive vulnerable users? Do such robots respect human dignity and difference?

First, while the intrinsic value of privacy is sometimes contested and seems to depend on context, privacy is in any case supported by other values such as personal autonomy. Most ethicists believe that privacy is important since it protects us from harm, and respects us as autonomous beings who want to have control over our lives as well as protect our personal sphere and human dignity. Information technologies are often seen as threatening privacy.[4] Personal robots may also be instrumental in supporting "surveillance capitalism," a new form of social and economic oppression based on the surveillance and manipulation of users, whose data are captured and sold.[5] As Carissa Véliz has argued, the captured data give tech companies the power to predict and influence our behavior.[6]

Do we want our private conversations and intimate moments to be captured by a robot? Where do the data go?

Using personal robots as home assistants renders us vulnerable to such a form of surveillance along with the threats it poses to privacy and power. In order to enable interaction via voice interface and the processing of human behavior, personal robots are usually equipped with cameras and microphones. They can record conversations and make videos. Do we want our private conversations and intimate moments to be captured by a robot? Where do the data go? And what is done with them? In the case of Jibo, the data went to the robot company. But who exactly has access to the data, and who sees and hears them? What happens to the data afterward? Do they leave the company? These worries are not about science-fiction scenarios. We already have smart dolls and smartphones that collect data and send them to the servers of the company for analysis. If this trend continues, we risk moving toward a total surveillance society in which surveillance not only happens on the street but in the kitchen and bedroom too. Not only governments (e.g., via the police, an intelligence service, etc.), but companies may watch you, or at least always have the possibility to do so; via robots, personal digital assistants, and other devices, information technology companies capture data about you, and can then use these data to predict and manipulate your behavior, or sell your data to other companies that wish to do this. And what if hackers get access to the data? Once technologies are connected to the internet, there is no such thing as 100 percent security.

Consider the example of Hello Barbie, a doll that records and analyzes what children say, and then talks back to them.[7] It sends the data to a ToyTalk control center for processing, using speech recognition. This enables the doll to have a "conversation" with children.[8] The children get a new "friend." But they are also under surveillance. And potentially the doll could be used for advertising, based on analysis of user data. Moreover, like any device that is Wi-Fi enabled and connected to the internet, hackers can get access to the personal data from the conversations and, through Hello Barbie, other personal data stored on devices in the home. Personal robots equipped with speech recognition and speech capabilities will create similar issues.

And robots not only collect audio data; your face and behavior may be recorded. Combined with AI in the form of facial recognition and other intelligent software, robots create new opportunities for surveillance, whether in the street and workplace, or the intimacy of home. If a robot can recognize your face, guess your intentions, and perhaps even read your emotions, any human who has access to these data and the results of the data analysis will have this information about you. Furthermore, because of the social interaction, such personal robots can provide extra insight into the personality of the robot's user via the user's behavior. As Ryan Calo has argued, the use of a robot is not the same as the use of a washing machine; how a person interacts with the robot, which program they use, and so on, says a lot about a person—and all of this is recorded and without legal protections.[9]

Second, mobile robots create an additional issue: safety. Imagine robots moving around the home with people there. A robot may hurt children when it bumps into them, for instance. It could be programmed to stop when a human is near. But this may limit its autonomous functioning. Hironori Matsuzaki and Gesa Lindemann call this the "autonomy-safety paradox."[10] Like the introduction of smarter industrial robots, more autonomous personal robots invite us to think about new safety concepts and their ethical aspects. It is also not clear where the responsibility lies when something goes wrong. And there is an ethical question related to humans and robots sharing the same space: Should humans always get priority—that is, should the robot always wait for humans—or should it sometimes take priority? How should the robots interact with humans (and vice versa) in particular situations, such as when both want to pass through the same door at the same time? And who takes the decisions about these issues: users or the robot company?

Deception, the Value of Real Human Relationships, Difference, and Gender and Racial Bias

The idea of using robot assistants for vulnerable users such as children and elderly people has sparked discussion about whether such robots can and should be used in care roles (see the next chapter), if users are deceived, and whether their dignity is respected. This concern applies not only to Wizard of Oz scenarios, in which users are deceived into thinking that the robot is autonomously operating whereas it not, but also to the idea of using personal robots that are actually autonomous and intelligent.

Consider autonomous personal robots that would be used by elderly people to provide care or companionship. Robert and Linda Sparrow have argued that such robots cannot meet the social and emotional needs of elderly persons, there is a risk of less human contact, and robots cannot provide genuine social interaction. They cannot provide real friendship, love, or concern. It is simulated. People may be misled into thinking that the robot cares. The Sparrows contend that this deception is bad because it misapprehends the world as it is, and because we want real love and companionship; having the belief is not enough: "What most of us want out of life is to be loved and cared for, and to have friends and companions, not merely to *believe* that we are loved and cared for, and to believe that we have friends and companions, when in fact these beliefs are false."[11] Even if our subjective well-being might be enhanced by the robot, what counts is real and objective love, companionship, and care. Robert Sparrow sketches a dystopian future in which robots keep people happy, but cannot provide respect and recognition. He imagines a windowless building in which elderly residents are watching screens and are cared for by robots—and no humans are around.[12] While one can philosophically question the way he frames the problem, and ask how "virtual" and "illusionary" robot care really is (the human-robot interaction is quite real), and whether deception is always and necessarily bad in care, it is clear that robots cannot offer the same connections as humans do, and that there is a danger of manipulation and disrespecting persons as well as their needs and human dignity.[13]

Similar concerns emerged in response to the idea of using robots as nannies for young children. Noel and Amanda Sharkey raise privacy and security issues (again, robots can record everything and be hacked), but also point to further problems. For example, should a robot be allowed to physically restrain a child, and, if so, in what circumstances? It probably should if the child wants to cross a dangerous road and risks running into car traffic, yet should it prevent the child from taking a cookie when that is forbidden by the parents? Where do we put the limits? And if robots take over "dirty" work such as changing diapers, children could miss out on building a good relationship with their (human) caregiver. Robotic care may endanger their mental and social development. The authors point again to the problem of deception too. Of course, children play with dolls and pretend during their play. But at the same time, children know that dolls are not really alive and it is play.

When a doll is replaced by an autonomous robot, however, a child might believe that they have a real relationship with the robot and become attached to it.[14]

This criticism is in line with Sherry Turkle, who acknowledges that children project when interacting with a (nonrobotic) doll, but maintains that this is different from the engagement with a personal robot, since in that case children engage with the robot "as if they face another person." But they are deceived; robots are not persons. The robot simulates. Instead of interaction with robots, Turkle argues, children need "relationships that will teach them real mutuality, caring, and empathy." More generally, personal robots raise the question of whether the quality and value of human relationships is in danger. Adults may also hope for companionship and expect it from the robot. Yet, Turkle asserts, robots cannot provide that. They offer us an "as if": they talk to us as if they understand us and care. This is not an authentic conversation, though. The problem is not only that the robots deceive us but also, in the end, that they make us forget what an authentic conversation is. She gives the example of the robot PARO, a robot in the shape of a baby seal sometimes used in homes for elderly people: the robot seems to listen and comfort an elderly person, but robots cannot empathize.[15] This is deception. And if we were to primarily interact with robots, would we forget how it is to have a real conversation, and what real companionship and friendship are like?

One could add a loosely Freudian perspective to Turkle's criticism of companion robots: we love robots in place of what we really want (love, friendship, good relationships, etc.).[16] The robot becomes a "fetish"—a term that is itself related to the Latin words for "artificial" and "to make." It is a human-made object that—in this context—comes to embody our desire for important human values. We imbue it with the magic of what we are *really* looking for and desire (in Sigmund Freud's view, sexual desire). According to this analysis, however, this is pathological. We better go for the real thing.

Talking about sexual life, the case of sex robots offers another example of how personal robots invite the criticism that what humans are, give, and do with one another cannot be replaced by robots and human-robot interaction. John Sullins doubts if sex robots can fulfill the physical and emotional needs of human beings instead of merely manipulating them. He contends that robots are not capable of creating an erotic relationship in which lovers encounter the other in their complexity, and learn and grow; instead, users of sex robots have the desire to "erase everything in the beloved that is not a complete reflection of the lover's preconceived notions of what he or she thinks they want out of a partner."[17] Again, the argument is that robots cannot provide the real thing. And perhaps sex robots are a way of avoiding the complexity and challenges of real human relationships. Once more they seem to threaten the development and exercise of good human-human relationships.

Not everyone agrees with this criticism; some are more open to the new possibilities offered by the technology. Are the mentioned criticisms too conventional and conservative? And is the problem in the robot or the user? Julie Carpenter has suggested that as people may get used to robots and culturally accommodate them, we accept them and then new norms may develop. Moreover, she has argued that ultimately the "rewards, benefits, burdens, and repercussions of emotional attachment do not rest on the robot or its design, but on the human who interacts with it."[18] Others have contended that sex robots can be a solution to sexual deprivation or are basically masturbation tools. (For an overview of arguments about sex robots, see John Danaher and Neil McArthur's *Robot Sex*.) In any case, it is worth noting that not only in ancient but also modern history there have always been complex and sometimes intimate relationships between humans and objects, such as erotic dolls, and some authors think that they are not necessarily fetishes or surrogates for something else.[19]

The discussion about sex robots is a good example of how personal robots, as social robots, make us reflect on *human* relationships. It makes us think about what good human relationships are and what good sex is. And even the concepts themselves are put into question. What are (good) human relationships? What is sex? What *is* "the

The discussion about sex robots is a good example of how personal robots, as social robots, make us reflect on *human* relationships. It makes us think about what good human relationships are.

real thing"? For instance, Carpenter's discussion about sex robots touches on notions of sexuality, attachment, love, romantic relationships, and theory about human-human relationships. The case of sex robots also makes us reflect on relationships between men and women, and more generally, gender issues. Kathleen Richardson, who launched a campaign against sex robots, has maintained that the use of sex robots projects an image of dealing with another in which the other—often a woman—is "turned into a thing." In this way, sex robots risk reinforcing asymmetrical power relations in which the human subjectivity of one of the partners is not recognized; empathy is "turned-off."[20] One could conclude that robots not only have direct ethical impacts such as breaches of privacy and security; at a deeper level, they are connected to the way we think about and do things with each other.

Against these objections one could submit that unfortunately, humans have not waited for personal robots to provide bad care, fail to empathize with other human beings, or treat others as things. One should not idealize what humans do to humans. Still, that humans do not always live up to high ethical standards is not a valid counterobjection since it does not undermine the arguments against (these uses of) personal robots as such; at most it may encourage fostering better human-human relationships in general, with and without robots. The deception charge, however, deserves some further discussion.

The arguments against personal robots sometimes make it seem as if deception is necessarily bad. But is this so? Much depends on how we define deception. If deception means deceptive speech, and if that is in turn defined as malicious lying, then most people would agree it is wrong. Yet if deception in speech is defined in terms of what Alistair Isaac and Will Bridewell describe as "a regular feature of everyday human interaction," encompassing "little white lies" and misdirection, then things look different. They assert that this kind of communication serves prosocial functions and genuinely social robots will "need that capacity to participate in the human market of deceit."[21] In other words, the reasoning is that if deception is part of what it means to function as a social entity, and if robots should become such entities, then they need to be able to deceive. This leaves unanswered, though, when (in what situations) little white lies are morally acceptable. Furthermore, the premise that we should build artificial social entities also needs more support. For example, why do we need robots that talk to us at all?

Second, deception could be defined as the creation of illusion. Is the making of illusions, which is arguably what the developers of such robots do, necessarily bad? For example, if an illusion is created in which the robot is a living being or friend, is this bad? Surely we do not have a problem with illusions when we go to the theater or a magic show. We are happy to be "deceived" during the show. Before and afterward, however—and perhaps during—we at the same time know that what is seen is not real. There is a so-called suspension of disbelief: people temporarily accept a show or story as reality in order to be entertained. Applied to personal robots, this would mean that people temporarily suspend disbelief during the interaction with the robot in order for the robot to do its "magic." Ethically speaking, then, one could demand from designers and those who offer the robot to users that users be made aware that the robot is creating illusions, that what goes on in the human-robot interaction is "as if" and make-believe. This is a challenge for designers and developers, but also for parents, care workers, and others who offer the robot to those they are supposed to take care of; it requires a kind of honesty about what the robot really is and can provide. This would go against much of the current advertising for personal robots, which are often sold as your "friend," supplying "companionship" and enabling "conversations," and so forth.

On the other hand, the line between reality and illusion is not as clear as it might at first glance seem. As said, the interaction with the robot is real, and the emotions on the part of the human are real during such human-robot "performances." The human-robot interaction does not take place in a fantasy world. Seen from this perspective, a strict Platonic distinction between illusion and reality seems difficult to maintain. Instead of there being two worlds (one real and one fake), things are more complicated. We could say that there is a performance by the robot and its users, and together they create something that is illusion and reality at the same time, not unlike what happens on the stage of a theater or magic show. Moreover, taking seriously the interactive aspect means recognizing that it is not just the robot or its creators that create illusions; users at least *co-create* the illusion as they participate in the robotic interaction and performance as spectators and potentially coactors. And as the term "suspension of disbelief" suggests, frequently users *like* to be deceived and deceive themselves. They want the magic. It seems therefore too simplistic to depict the designers and caregivers as necessarily and wholly guilty, and frame the user as entirely innocent. At least there seems to be a crucial difference between, for example, adults and young children; most adults may be far less innocent co-creators of illusion when they use personal robots (they temporarily accept the illusion, but know that it is an illusion), whereas additional ethical concern seems in order with regard to vulnerable users such as young children or, say, persons with cognitive impairments who might not know at all that it is an illusion. The ethics of creating illusions, and therefore the ethics of personal robots, is more complex than presented here.[22]

Beyond issues regarding deception and intimate human relationships, it is questionable whether the use of personal social robots sufficiently respects difference—

cultural and personal differences. If robots are not just artifacts but also are always linked to their social and cultural environments, then this has ethical implications. Robots are *developed* in a particular society and part of the world, such as the United States or Japan, and the way they appear and behave is likely to be influenced by the values as well as form of life of that society and culture. This may be in tension with the society, culture, and values of the user and the environment in which the robot is *used*. For example, a robot and AI designed to function in an individualist culture—say, by having learned from Western internet texts and conversations—may not always work well in a more communal or collectivist cultural environment. Moreover, the design and development of robots may assume a kind of universal standard user, but we are all different people, with potentially different needs and expectations, different bodies and genders, and different personal histories. Consider, for example, personal robots whose interaction with humans relies almost entirely on voice interface, which cannot be used by users with hearing impairment. Or robots that assume a binary gender distribution in their communications, talking as if there are no transgender people or others who do not fit these categories. There are also different expectations about what good care is: some people want minimal social contact and to be left alone in some situations, others want to talk a lot and expect empathy, and so on. How can and will personal robotics deal with this? AI could be used to have the robot's intelligence adapt to the user and its culture. But the potential for friction is likely to remain. Humans can learn to deal with difference and exercise respect for others who are different than them. Can machines provide that kind of respect to humans given that they cannot even care about anything at all? Perhaps they could *simulate* respect for difference. Would that be enough? Again, his raises objections concerning deception along with violations of human needs and dignity.

Finally, a related societal and cultural issue is that personal robots may have behaviors, meanings, and wider implications that create, sustain, or increase bias, such as gender or racial bias, and more generally impact justice and equality in a specific society. I already mentioned that sex robots raise gender issues. But other robots may be problematic in terms of gender too. Technological artifacts such as robots may be gendered, in the sense that if they are connected to normatively relevant gender meanings available in a particular culture and society, or what I have called "gender grammars" and "gender games."[23] For example, a cleaning or household robot in the shape of a woman is problematic since it perpetuates perceptions and practices of categorizing women as belonging in these roles. More generally, robots are linked to our cultural expectations in their roles as servant, friend, slave, and so on, and these roles are "loaded with user stereotypes," as Carpenter and colleagues put it.[24] Even if this is usually not intended by the designers, robots have this effect since there is always a link to society; in this sense, robots are never mere things.

The gendering of robots is thus problematic since it may "perpetuate aspects of certain human-human roles and the ideologies that go with them."[25] This is also true for discrimination and racism. As Ruha Benjamin has argued, automation technologies such as robots and AI may unintentionally deepen discrimination and have racist effects.[26] In a US context, for instance, this can mean that such technologies risk reinforcing a history of oppression and promoting white supremacy. While not about robotics as such, machine learning used in a beauty contest with AI judges can lead to racial bias. Other famous illustrations include photo-tagging software that classified dark-skinned users as gorillas, or predictive crime algorithms that are biased against people from Black neighborhoods. But Benjamin also points to some other ways in which robots can be problematic in terms of race and related societal concerns: the history of robotics with its desire for personal slaves, the use of robots by police for surveillance and lethal force against particular racial populations (an extremely relevant theme as I write this book), sensors that only recognize a small spectrum of skin tones, and inequalities in the tech labor force along racial and gender lines. Consider low-paid manual laborers in the Global

South, often women, doing work for white creatives and entrepreneurs positioned at the top of the hierarchy. Although usually there is no racist intention on the part of the designers and developers of the robots, there is a significant risk that robots perpetuate a history of domination and racism, continuing what I would call, in analogy to "gender games," particular "race games."[27] A particular society already has specific ways of dealing with race, and these wider social and cultural contexts, including its (frequently unwritten) rules and the problems they raise, are mirrored, performed, and maintained in the use of technologies such as robots and AI. In this sense, robots can be racist, as Benjamin puts it, or as I would say, the performances and games in which robots play a role can be racist, and robots can continue and promote these games.

As is clear from the examples, this analysis in terms of gender and race games is not only applicable to personal robotics but robots in general too. But given their social function(s), personal robots, particularly when they have a humanlike appearance, require special ethical attention when it comes to issues such as gender and race. The problem with human-looking robots is not only that they look human (and hence raise the deception problem); the problem is that they look like *particular* (categories of) humans, and may have potentially harmful effects on particular (categories of) humans as they interact with (particular) humans and have wider societal effects.

To conclude, personal robots (as social robots) raise crucial metaphysical, ethical, societal, and existential questions. The discussion easily takes us beyond "ethics" if that domain is narrowly defined as being about what is right and wrong; it also makes us think about what is real, what we value and want, problems in society, and what it means to be human. As Carpenter observes, experiencing and thinking about such robots, we may gain "insights about what it means to be human or not human," and it "may make us reflect on ourselves and what we really want from each other."[28]

Ethics of Personal Robotics beyond Humans

The previous discussion assumed that in robot ethics, we should only be concerned about humans. But there are further ethical issues related to the fact that personal robots operate in environments where there are not only humans but also *animals*. What are the ethical implications of such robots for pets, for example? How do, for instance, cats and dogs react to robots, and vice versa, and what is an ethical way for robots to deal with them? There are, say, safety issues. One could demand that robots that move around the home should not harm animals. Yet given the social and conversational character of robots, there is the broader question, How do and should robots intervene in,

The discussion easily
takes us beyond "ethics";
it also makes us think
about what is real,
what we value and want,
problems in society,
and what it means to
be human.

and influence, the social fabric of relationships between humans and animals? What will and should be their place and role? Will the place and role of pets change when robots are introduced in the home? What will the robot-animal interactions look like? How should we shape these social interactions and mixed human-nonhuman networks? Assuming that ethics should take into consideration not only the interests, needs, and value of humans but those of animals too, these are outstanding challenges for personal robotics—one that are seldom addressed.

Moreover, beyond the impact on our relationships with animals, there is the question of the *environmental* impact of these robots. Not only industrial robots, but personal robots will impact the natural environment because of the way they are produced, used, and handled after use. For example, the ethics of personal robotics should ask, When the robot breaks down (or is shut down by the company), what happens to the hardware? Does it end up as waste? Will it be recycled? How should it be produced? How much energy does that take? What materials should be used? What is the carbon impact? And what are the responsibilities concerning these issues on the part of robot designers, companies, users, and public authorities? The case of smartphones, the personal devices that most of us already have, is not promising: they have brief life cycles and are seldom recycled, and their production, operation, and infrastructures involve a lot of energy consumption (e.g., running the servers), CO_2 emissions, and mining of rare minerals. Will the same happen with our robot "friends" when the company that makes or sells them encourages us to buy a new one every two years? And what will be the carbon footprint of the infrastructure needed for the energy and the AI that powers all the robots?

In the concluding chapter, I will return to these questions about a robot ethics that goes beyond the human.

4

CARE ROBOTS, EXPERTISE, AND THE QUALITY OF HEALTH CARE

In the 2012 science-fiction film *Robot & Frank*, an aging man with dementia gets a robot from his son, who is tired of making regular visits to his father. The robot takes on domestic tasks and provides therapeutic care, monitoring Frank's daily routines and food, but also encouraging gardening to maintain Frank's cognitive abilities. While Frank at first doesn't like the robot, he soon appreciates the robot and uses it for his own plans. The robot becomes a friend, and no longer a servant or slave. His daughter, however, thinks that the machine is ethically objectionable. There is even a social movement against it. Are care robots a way to abandon the elderly? For example, if remote control and monitoring is possible, will people be less likely to visit their elderly parents? And is it OK to become totally dependent on the technology? Can the robot help with chores, without people becoming totally dependent on it? Frank's daughter says when she agrees that the robot should be used. But if we see what has happened with cell phones, people may well get *very* reliant on the technology.

This scenario is not too far-fetched. The Japanese government thinks that robots are a way to cope with the growing elderly population.[1] A related rationale for using care robots is saving personnel costs. And there are already robots being tested and used in health care, not just in Japan, but in other parts of the world. For example, the robot PARO, made by the Japanese company AIST, is marketed as a "therapeutic robot" that can and has been used in hospitals and care facilities to relax patients as well as stimulate social interaction between patients and between patients and caregivers. It looks like a baby seal, and responds to stroking and voice. "By interaction with people, PARO responds as if it is alive, moving its head and legs, making sounds, and showing your preferred behavior. PARO also imitates the voice of a real baby harp seal."[2] It may be true that social interaction is stimulated. But could there be something wrong with using such robots in this way? Is it deceptive, as asked in the previous chapter? Is it humiliating? Is it infantilizing people? Or should the therapeutic benefits be stressed? What are the ethical issues raised by using robots in health care?

Robots in Health Care

The PARO robot is an illustration of how robots can be used in the care of the elderly. Robots could be used in nursing care in general too. For instance, in a hospital environment they can be used to get medical supplies and food, deliver medicines, or move patients. There are already robots that help nurses by doing noncare tasks such as bringing supplies and medicines, like the robot Moxi in the United States.[3] Or they can play a role in so-called telemedicine; patients living in remote areas, people in urgent need of seeing a doctor, or people who receive health care at home can get a diagnosis through a robot that is remotely controlled. Robots could also (help to) monitor people who stay at home and possibly intervene when something goes wrong. For elderly people, say, a robot could check for signs of a fall. Next to assistance and monitoring, it is claimed that some robots can provide companionship (see the discussion in the previous chapter).

But robots can also be used in other health care areas. In robot-assisted surgery, the surgeon performs the surgery remotely through computer control or by using a telemanipulator. In this case, the surgeon does not have direct contact with the patient's body; the robot does the actual surgery. The technology is mainly used for so-called minimally invasive surgery, which means that only a small incision is needed. For example, in the case of the da Vinci surgical robot, instruments and a 3D-camera are mounted on robotic arms and controlled by the surgeon, who controls the arms (all at once) by means of a console with a monitor that gives them a good view of what happens inside the patient. For the surgeon, this requires specific training and learning new skills, but once learned, the system means increased control and thus more precise surgery. For the patient, minimally invasive and more precise surgery means less trauma and faster recovery.[4] Today the use of these robots is often accepted by patients, although there are some ethical issues. For instance, patients need to be sufficiently informed, the cost of the technology is relatively high, and when something goes wrong, it is not clear how to distribute the legal and moral responsibility since there are many parties involved, including at least the surgeon, hospital, and manufacturer (similarly, on responsibility for self-driving cars, see the following chapter).[5]

The example of surgical robots shows that robots in health care do not necessarily take the form of humanoid robots that operate independently—say, the image we get from science-fiction films. Of course, there *are* robots that look like humans or animals. Consider again PARO. But there are, for instance, exoskeleton robots. The robot HAL is used in rehabilitation and helps people to walk again.[6] Moreover, robots are part of larger sociotechnical systems. There is not only a "dyadic interaction between a human and a robot."[7] Robots impact health care systems as a whole, such as by changing the role of health care staff, and having implications for education and expertise.

More controversial than surgical robots is the use of socially assistive robots in mental health care, where they have been used, for instance, in dementia care and therapy for autistic children. In the latter case, the robot can be used to replace the therapist, play a mediating role, or assist the therapist. The first role has been tested, for example, in the European project DREAM, which used a robot with supervised autonomy; when not remotely operated (Wizard of Oz), the robot operates autonomously for some time, but the therapist is still present.[8] This raises all kinds of questions regarding the appearance and behavior of the robot (e.g., is it acceptable to have the robot, such as the NAO robot, look like a human?), responsibility, trust, and privacy.[9] And as with care robots, the worry is that the robots are used simply to save money with no clear benefit for the patient. In this case, these issues were investigated and discussed as part of the research project.[10] But often this ethical component is missing in robot development.

Ethical Issues

In general, robots in health care raise a number of ethical issues. Let me provide an overview, with an emphasis on intelligent and autonomous robots in care tasks in the context of nursing and elderly care.

First, like all networked technologies, there are again issues concerning privacy, data protection, and surveillance. Which data are collected by the robot, how are they stored, who has access to them, who owns them, and what happens with the data now and in the future? Whether the robots are used at home (see also the previous chapter) or in care facilities such as hospitals and homes for elderly people, the technology enables constant surveillance. Is this ethically acceptable? And what if people cannot yet or can no longer give their consent?

Second, as with industrial robots, there is the question regarding the implications for labor. If and to the extent that the robot takes over tasks from humans, fewer humans may be necessary; the result is unemployment. And particular (categories of) humans working in care roles may be seen as no longer necessary and hence particularly affected, even if unintentionally, such as women or migrants. As indicated, one of the main reasons given for the introduction of robots is to save personnel costs. But technological solutions are not the only or necessarily best way to tackle demographic challenges. One could employ more human care workers. There is the suspicion that visions of robotics in health care have less to do with improving the quality of care, and more to do with justifying public and private investments in the tech sector, and the replacement of human care workers in order to reduce labor costs (and in some cases, not to employ immigrant workers; it is frequently suggested that this plays a role in the Japanese context). These are thus not just economic problems or labor problems but also related to issues concerning justice and equality.

Third, next to these "macro" problems regarding replacement and its socioeconomic impact given the already-mentioned systemic embeddedness of robots, there are more "micro" and "meso" issues that arise from the autonomy of the robot. What is the impact on organizations, practices, and interactions? Replacement is again a key concern here. What, exactly, will be the role of the robots? Do they take over entire jobs or just specific tasks? Do these tasks involve collaboration with humans? What is the resultant role of the nurse, doctor, and surgeon? What does this mean for the organization? What is the impact on partners, relatives, and friends? There is the worry that the use of care robots leads to less human contact.[11]

Furthermore, who is responsible when something goes wrong? There is a "responsibility gap" ' in the sense that the robot gets more autonomous and is given more tasks, but robots lack the capacity of moral agency.[12] How are humans to be held responsible if they do not have sufficient control? How can the attribution of responsibility—to humans—be secured if the robot is not supervised? (See also the next chapter.) This problem can be solved by having

Technological solutions are not the only or necessarily best way to tackle demographic challenges.

supervised robots. But then why use robots at all (if the aim is reducing personnel cost), and to what extent is the supervisor responsible when the robot does something wrong given that it is developed by others?

Beyond the cost issue, one reason why we might need robots is to make specific tasks for care workers easier. Robots are then not used for replacement but instead assistance and augmentation. Consider a nurse who has to lift a patient; a robot can be helpful in this case since it prevents back problems on the part of the nurse. Consider again the case of robotic surgery or indeed supervised mental care. Such cases of assistance and augmentation seem to be a lot less problematic than, say, visions of a robot that distributes medicines to patients or takes over the entire nursing job. The ethical discussion thus depends a lot on which tasks and roles the care robots (would) get.

Fourth, there are issues that concern the human-robot interaction, which have been mentioned already in the previous chapter and introduction, such as deception, trust, and freedom. If the robot pretends to be an animal or human, is this a case of deception, and is this wrong per se? Consider again the case of PARO, and more generally, the discussion initiated by the Sparrows and Sharkeys.[13] Are people cared for by personal robots living in illusion? Are they deceived? And can the robot be trusted by patients and care workers? Is "trust" the right word? Or is it just used as a nicer-sounding term than "acceptance," which is what robotics researchers and the industry tend to care about? And should robots be allowed to restrict the freedom of, say, an elderly person with dementia or person who is suicidal?

Robots in care contexts may raise gender issues as well. I already introduced this issue in the previous chapter and mentioned the potential gender aspect of unemployment risk, but there are also gender issues at the nexus where micro and macro meet—that is, in the human-robot interaction as linked to cultural expectations and norms. For example, even today nursing is still often perceived as an occupation for women; care robots with a female voice and/or appearance may reinforce this stereotype. The discussion about robots in care thus reveals assumptions—including stereotypes—about women, nurses, and (health) care.

Fifth, related to the concern that there is less human contact is the argument that robots cannot provide a kind of "warm" care—that is, cannot provide a kind of care that responds to the social and emotional needs of patients. The idea is that patients, elderly people, and so on, not only need "cold" technical care in the sense of medical procedures or medicines but also want to have a chat, have their emotional needs recognized, have their dignity respected, and so on. They need meaningful conversation and emotional support. Robots cannot provide this since they lack emotions and consciousness. Therefore, it is concluded, robots should not be used in health care. Against this argument it has been remarked that current health care is already rather cold and even that it should be cold, since, first, too much emotional involvement would make the work of care workers impossible and, second, not all patients want warm care; some want a more distant kind of care. For example, it is said that some would prefer a machine to wash them rather than a human being because it is less shameful and more dignified. A defender of "warm" care could reply that the fact that current health care is cold is not a valid argument against cold care as such; clearly there are also problems with the current health care, but this alone does not render the use of robots acceptable. Moreover, while it is true that too much involvement may be problematic for care workers, this does not say anything against including *some* degree of warmth—that is, some response to emotional needs and some social dimension. And while some people might prefer cold care, it is important to cater to people who want a different kind of care. And what if the people who prefer cold care are in a situation where they no longer have family (or their family no longer visits regularly) and the care workers are the only human beings they meet?

Sixth, when robots are used for adults, such as elderly people, there is the concern that this might not respect the dignity of people and instead infantilize them. Consider the robot PARO again; giving this robot to elderly people

When robots are used for adults, such as elderly people, there is the concern that this might not respect the dignity of people and instead infantilize them.

with dementia could be done in a way that treats them as children. Or worse, they could be treated as things rather than people. One of the problems concerning the treatment of people with dementia is what Tom Kitwood has called "objectification": "treating a person as if they were a lump of dead matter."[14] More generally, it is important that robots do not threaten the dignity of people—and of course that human dignity is respected *in general* in health care, with or without robots. The Sharkeys offer a chilling story about an elderly person who does not get a bedpan to urinate in because, as the nurse tells his wife, the sheets will be changed anyway. The issue concerning people with dementia also raises the question of the kind of image we have of those who need care, and how to deal with vulnerable users while taking into account their diverse conditions and needs (next to the needs of the care workers).

These issues concerning deception, trust, warm care, and human dignity are related to more general questions that should be asked with regard to the introduction of robots in health care. What is good care? And can robots provide it?

What Is Good Care?

The Sparrows' argument against using socially assistive robots in health care was not only about deception but rather, more generally, the quality of care. One could contend that what really matters is not so much (the absence of) deception, trust, or warm care as such but instead *good* care. In particular, against the notion that it is acceptable to deceive people if that makes them feel better, the Sparrows maintained that what counts is not just that patients *feel* good, subjectively, but also that objective criteria of good care are met.[15] But what is good care?

Let us start with the assertion that the feeling and experience of care is not the *only* thing that counts. A device that can be used to argue for this view (and hence further support the point that the Sparrows already made) is what I have called the "care experience machine," modeled on the thought experiment the "experience machine" from Robert Nozick.[16] Nozick writes:

> Suppose there were an experience machine
> that would give you any experience you desired.
> Superduper neuropsychologists could stimulate
> your brain so that you would think and feel you
> were writing a great novel, or making a friend, or
> reading an interesting book. All the time you would
> be floating in a tank, with electrodes attached to
> your brain. Should you plug into this machine for life,
> preprogramming your life's desires? . . . Would you
> plug in? What else can matter to us, other than how
> our lives feel from the inside?[17]

Suppose, then, that if robots could give patients the experience of receiving good care, would such "virtual" care, to use the Sparrows' term, be ethically acceptable? The thought experiment is of course designed to make you answer, "No, subjectively good care is not enough. We need real care and good care." The point is not and should not be that the experience of care is not relevant at all. One could still claim that experience counts, but that necessarily objective criteria of good care should be met at the same time. Subjective criteria that are at once objective criteria constitute necessary conditions, but neither condition is sufficient for good care on its own. So what is *objectively* good care?

Objective criteria for good care can be derived from a number of sources. The Sharkeys propose to assess the question regarding care robots by using human rights and shared human values.[18] Consider rights to health and well-being, privacy, and freedom from degrading treatment (for the latter, Article 5 of the Universal Declaration of Human Rights can be used), or values such as welfare, privacy and consent, and accountability. The Sharkeys emphasize human welfare. There are also general ethical principles developed for bioethics and widely used in medical ethics. Tom Beauchamp and James Childress have famously proposed the principles of autonomy, beneficence, nonmaleficence, and justice.[19] These are meant to guide all health care workers and could be applied to the ethics of using robots in health care. These rights and principles, however, are formulated at a general level and in practice they are mostly negatively formulated—that is, as prohibitions; it is still an open question what kind of positive vision of care should be the aim. For example, the bioethics principles of not harming patients and doing good gives little guidance for the concrete questions of whether robots should be introduced in health care, and if so, for which tasks and roles, and for working with which kinds of patients. The bioethics principles also fail to put sufficient emphasis on the social dimension of health care.

By contrast, the Sparrows and assume in their paper that good-quality health care must include (real) emotional care and social relationships, given the social and emotional needs of persons.[20] Both the Sparrows and Sharkeys hold the view that doing routine care tasks often provide the opportunity for social interaction—for elderly, but, for instance, for children (see also the previous chapter)—and hence contributes to an important human need and right.[21]

In line with this insight about the social dimension of care, but referring to other general principles and normative theory, I have proposed that we can use Martha Nussbaum's version of the capability approach to evaluate robots in health care.[22] The capability approach is designed to give guidance on human development, but it is also relevant to thinking more generally about quality of life, human dignity, and justice. Nussbaum's list includes the following "central human capabilities":

1. Life: "being able to live to the end of a human life of normal length; not dying prematurely, or before one's life is so reduced as to be not worth living."[23]

2. Bodily health (including nourishment and shelter).

3. Bodily integrity: free movement, freedom from sexual assault and violence, and having opportunities for sexual satisfaction.

4. Being able to use your senses, imagination, and thought: experiencing and producing culture, and freedom of expression and religion.

5. Emotions: being able to have attachments to things and people.

6. Practical reason: being able to form a conception of the good and engage in critical reflection about the planning of one's life.

7. Affiliation: being able to live with and toward others, imagine the other, and respect the other.

8. Other species: being able to live with concern for animals, plants, and nature.

9. Play: being able to laugh, play, and enjoy recreational activities.

10. Control over one's environment: political choice and participation, being able to hold property, and being able to work as a human being in mutual recognition.[24]

Applied to good health care, using these criteria implies that it is necessary but not sufficient if care helps to restore the health of patients or keep elderly people alive; it should also contribute to their social and emotional well-being, lead to respect for others, give people control over their environment, take into account people's own plans, and so on. The question regarding robots should then be discussed in the context of such a broader vision of health care as being aimed at the enhancement, maintenance, and restoration of central human capabilities. For example, does the technology really enhance or maintain the capability of affiliation with others, or does it diminish social contact, as the Sparrows suspected? Does the introduction of care robots respect people's capacity for practical reason or, for instance, infantilize them—thus threatening the dignity of people (e.g., elderly persons)? And if we consider the question regarding justice, will these benefits only be available to rich people in technologically advanced societies?[25] While perhaps one should not ask *too much* of care workers, such questions can help us to think about what kinds of care we want, which in turn is a good ethical anchoring point for questions about care robots. This approach has inspired others to propose a capability approach to robotic care.[26] The capability approach may also offer a way to broaden definitions of health and well-being from bodily health to the mental and social dimensions of health—and preferably to go beyond such dualistic categories altogether. This is important for the ethics of robotic care and care in general.

Another approach, which looks at the question of good care by putting it in the context of the quality of human lives in general, is to ask the ancient Aristotelian philosophical question concerning *the good life* (*eudaimonia*) and derive some normative guidance from this. For example, what does human flourishing mean for ill or elderly people? Could Aristotle's notion of friendship guide care by family and friends, or is that too demanding, and designed for relationships between equals, unsuitable for thinking about caregivers and dependents? What does companionship mean in the context of health care? What is virtuous health care, and what could the role of robots be in such care?[27] Sparrow also refers to Aristotelian virtue and the good life when arguing for elderly care that meets the ethical requirement of respect as well as the "objective good" of recognition.[28]

In a related vein, I have asked what virtuous care *work* is.[29] I have argued that such work is not only social but involves skilled and bodily engagement too—a kind of craftspersonship, which relies on tacit knowledge and know-how. Moreover, it is directed at excellence or virtue.[30]

Care workers are and should be virtuous craftspersons, if and insofar as they skillfully, bodily, and socially engage with people and things in order to reach care excellence. For example, nurses can be considered to do exactly this kind of work (or at least should be doing this kind of work) as they engage with patients and all kinds of things in a skilled way, involving the learning of know-how, next to "know-that" (theoretical knowledge). Can robots achieve such a craftspersonship? Or should the robot only be used as an aid to human care craft?

Based on the work of Stuart and Hubert Dreyfus, one could question if care *expertise* and the related ethical know-how, both of which involve a kind of embodied and intuitive know-how, can be transferred to robots.[31] Consider again robotic surgery; here the robot is still remotely operated by the surgeon, who has to acquire particular skills and embodied expertise for teleoperating the robot. In and as part of their work, understood as a craft, the surgeon will have developed a sense of what is ethically acceptable and good practice. In this sense, craftspersonship is retained, or at least a new kind of craft is learned: the craft of operating mediated by the robot. But can the robot take over completely, achieving the technical and ethical expertise of the human? These accounts of care work and care expertise thus raise the question of whether care work, understood as craftspersonship and expertise, can be done by robots.[32] There is at least a danger that skilled engagement and craftspersonship are eroded when the robot replaces the human care worker.

One problem with both the capability and good life approaches, though, is that their Aristotelian emphasis on human flourishing might be seen as privileging those stages in life when we are adult, independent, not ill, not elderly, and have our full capacities; it can be questioned how suitable it is for those stages in life when there is "a natural decline in capacities."[33] In response, one can adapt the good life approach (e.g., I talk about "the best possible life") and question the image of elderly persons as people in decline, or use another approach.[34]

Based on my review of the discussion about robots in health care and for more practical purposes, I have proposed some general criteria for good care that directly respond to some of the most important issues identified in the previous section and include some of the general principles outlined in this section, but without direct references to the Aristotelian framework or the ideal of craftspersonship. What I call "a normative ideal of good care" includes the following working criteria:

Good care attempts to restore, maintain, and improve the health of persons.

Good care is practiced within the boundaries set by bioethical principles and professional codes of ethics.

Good care involves a significant amount of human contact.

Good care not only means physical care but also has psychological and relational, such as emotional, dimensions.

Good care is not only professional care but should involve relatives, friends, and loved ones to a significant degree too.

Good care is not (only) to be experienced as a burden but also can be (experienced as) meaningful and valuable.

Good care involves skilled engagement with the patient (know-how), next to more formal forms of expertise (know-that).

Good care requires an organizational context in which there are limits to the division of labor so as not to make the previous criteria impossible to meet.

Good care involves an organizational context in which financial-economic considerations are not the only or even main criterion in the organization of care.

Good care requires the patient to accept some degree of vulnerability and dependency on others.[35]

The latter criterion concerns care receivers; going beyond discussions centered on caregivers and their duties,

Good care not only means physical care but also has psychological and relational, such as emotional, dimensions.

the assumption is that good care is not just a matter of what caregivers do but also depends on the care receivers, who share at least some of the responsibility for good care (e.g., self-care) and (should) have different options to cope with the new technologies.

Of course these criteria for (objectively) good care can and should be used to assess health care without robots too. Furthermore, in my view these lists should be used in addition to taking into account people's subjective experience, which is certainly not the only thing that matters—see again the thought experiment—but should at the very least *also* matter. Just as it is not good to have subjectively good care without objective benefit, it would be wrong to have experts tell people that they receive good care without them *feeling* that it is good. Both dimensions are necessary for good care.

This subjective criterion, together with the demand to respect human dignity, remind us that one should not forget to actually *ask* the people involved, such as care workers, care receivers, and other stakeholders. For example, in the case of robots for autistic children, we have done a survey and at a later stage organized a workshop with stakeholders.[36] The rationale is that instead of leaving the decisions only to experts, it would be better to have a deliberative process that gives people a say in how the technology is used (and preferably, if it should be used at all). For instance, one could involve elderly persons to test the ethics of the use of robots in aged care.[37] This approach also fits the concept of responsible research and innovation, which concerns at least two aspects: one is about doing ethics early in the process of development and innovation rather than afterward, and another is about involving stakeholders.

The first aspect is important since engineers, computer scientists, and others involved in the development of robots should carry co-responsibility for its use and consequences. Sometimes this approach is formulated in terms of values; developed by Batya Friedman and others, so-called value sensitive design takes human values into account throughout the design process.[38] The concept has been applied to robotics too. Aimee van Wynsberghe, for example, has argued that ethics ought to be included in the design process of care robots, promoting the values and dignity of patients.[39]

The second aspect stresses the co-responsibility of other stakeholders and is often linked to the ideas of participatory design as tied to participatory democracy. It is meant to close the gap between technology and society:

> Responsible Research and Innovation is a transparent, interactive process by which societal actors and innovators become mutually responsive to each other with a view on the (ethical) acceptability, sustainability and societal desirability of the

> innovation process and its marketable products (in order to allow a proper embedding of scientific and technological advances in our society).[40]

Again, this concept can and has been applied to robotics. In light of this responsible research and innovation vision, Bernd Stahl and I have argued for an ethics of health care robotics that next to an analysis of ethical issues, includes forms of reflection, dialogue, and experiment that come close to innovation practices and contexts of use.[41] This is another meaning of "good care," and arguably, good robotics and good *technology*.

To conclude, thinking about robots in care not only addresses specific ethical issues raised by care robots but also makes us think about what good care is, and how to make sure that relevant values and stakeholders are integrated in care along with the development and use of care technology.

5

SELF-DRIVING CARS, MORAL AGENCY, AND RESPONSIBILITY

Imagine a self-driving car drives at high speed through a narrow lane. Children are playing on the street. The car has two options: either it avoids the children and drives into a wall, probably killing the sole human passenger, or it continues its path and brakes, but probably too late to save the life of the children. What should the car do? What will cars do? How should the car be programmed?

This thought experiment is an example of a so-called trolley dilemma. A runway trolley is about to drive over five people tied to a track. You are standing by the track and can pull a lever that redirects the trolley onto another track, where one person is tied up. Do you pull the lever? If you do nothing, five people will be killed. If you pull the lever, one person will be killed. This type of dilemma is often used to make people think about what are perceived as the moral dilemmas raised by self-driving cars. The idea is that such data could then help machines decide. For instance, the Moral Machine online platform has gathered millions of decisions from users worldwide about their moral preferences in cases when a driver must choose "the lesser of two evils."[1] People were asked if a self-driving car should prioritize humans over pets, passengers over pedestrians, women over men, and so on. Interestingly, there are cross-cultural differences with regard to the choices made.[2] Some (collectivist) cultures such as Japan and China, say, were less likely to spare the young over the old, whereas other (individualist) cultures such as the United Kingdom and United States were more likely to spare the young. This experiment thus not only offers a way to approach the ethics of machines but also raises the more general question of how to take into account cultural differences in robotics and automation.

Figure 3 shows an example of a trolley dilemma situation: Should the car continue its course and kill five pedestrians, or divert its course and kill one?

Applying the trolley dilemma to the case of self-driving cars may not be the best way of thinking about the ethics of self-driving cars; luckily, we rarely encounter such situations in traffic, or the challenges may be more complicated and not involve binary choices, and this problem definition reflects a specific normative approach to ethics (consequentialism, and in particular utilitarianism). There is discussion in the literature about the extent to which

Figure 3 Example of a trolley dilemma presented to Moral Machine participants.

trolley dilemmas represent the actual ethical challenges.[3] Nevertheless, trolley dilemmas are often used as an illustration of the idea that when robots get more autonomous, we have to think about the question of whether or not to give them some kind of morality (if that can be avoided at all), and if so, what kind of morality.

Moreover, autonomous robots raise questions regarding moral responsibility. Consider the self-driving car again.

In March 2018, a self-driving Uber car killed a pedestrian in Tempe, Arizona. There was an operator in the car, but at the time of the accident the car was in autonomous mode. The pedestrian was walking outside the crosswalk. The Volvo SUV did not slow down as it approached the woman.[4] This is not the only fatal crash reported. In 2016, for instance, a Tesla Model S car in autopilot mode failed to detect a large truck and trailer crossing the highway, and hit the trailer, killing the Tesla driver. To many observers, such accidents show not only the limitations of present-day technological development (currently it doesn't look like the cars are ready to participate in traffic) and the need for regulation; they raise challenges with regard to the attribution of responsibility. Consider the Uber case. Who is responsible for the accident? The car cannot take responsibility. But the human parties involved can all potentially be responsible: the company Uber, which employs a car that is not ready for the road yet; the car manufacturer Volvo, which failed to develop a safe car; the operator in the car who did not react on time to stop the vehicle; the pedestrian who was not walking inside the crosswalk; and the regulators (e.g., the state of Arizona) that allowed this car to be tested on the road. How are we to attribute and distribute responsibility given that the car was driving autonomously and so many parties were involved? How are we to attribute responsibility in all kinds of autonomous robot cases, and how are we to deal with this issue as a profession (e.g., engineers), company, and society—preferably proactively before accidents happen?

Some Questions concerning Autonomous Robots

As the Uber accident illustrates, self-driving cars are not entirely science fiction. They are being tested on the road, and car manufacturers are developing them. For example, Tesla, BMW, and Mercedes already test autonomous vehicles. Many of these cars are not fully autonomous yet, but things are moving in that direction. And cars are not the only autonomous and intelligent robots around. Consider again autonomous robots in homes and hospitals. What if they harm people? How can this be avoided? And should they actively protect humans from harm? What if they have to make ethical choices? Do they have the capacity to make such choices? Moreover, some robots are developed in order to kill (see chapter 7 on military robots). If they choose their target autonomously, could they do so in an ethical way (assuming, for the sake of argument, that we allow such robots to kill at all)? What kind of ethics should they use? Can robots have an ethics at all? With regard to autonomous robots in general, the question is if they need some kind of morality, and if this is possible (if we can and should have "moral machines"). Can they have moral agency? What is moral agency? And can robots be responsible? Who or what is and should be responsible if something goes wrong?

Moral Machines? Can Robots Be Moral Agents?

Let us start with the question regarding moral agency and moral machines. Some researchers think that robots can be moral agents and try to develop moral machines. For instance, Michael and Susan Anderson do "machine ethics" in the sense that they aim at creating ethical agents that make decisions in the way that humans do. They think that "ethics can be made computable," and we should give machines principles and let them reason in a rational way, thereby avoiding human beings who "get carried away by their emotions."[5] Alan Winfield, a roboticist, also believes that we can create a moral machine, which he defines in one talk as "a robot capable of deciding or moderating its actions on the basis of ethical rules."[6] What kind of rules? Winfield gives the example of a robot that prevents a human from falling into a hole by following the rule that humans should not come to harm. Winfield refers to the first law of Isaac Asimov's laws of robotics, which Asimov proposed in his robot science-fiction stories. The first law is that "a robot may not injure a human being or, through inaction, allow a human being to come to harm."[7] In terms of normative moral theory, the robot would follow consequentialist ethics, preventing bad consequences for humans (e.g., the harm of someone falling into a hole). For this purpose, Winfield argues, we do not need a sentient robot but rather one that follows ethical rules, and in order to be able to implement these rules, it must be able to have a model of itself and its environment (including others in that environment) so as to predict the consequences of its own actions as well as those of others. A robot with such an "internal model" would be able to generate what-if hypotheses such as, What if I carry out action x, and which of several possible next actions should I choose?[8] In the hole example, the robot needs to be able to predict what the human is going to do (fall into a hole) and figure out what it has to do in order to prevent that. The rationale for developing such ethical robots is that once we have more autonomous and intelligent robots that can take on all kinds of tasks and respond to their environment, safe robots are no longer enough. As Wendell Wallach and Colin Allen write in their book *Moral Machines*,

> If multipurpose machines are to be trusted, operating untethered from their designers or owners and programmed to respond flexibly in real or virtual world environments, there must be confidence that their behavior satisfies appropriate norms. This goes beyond traditional product safety. . . . [I]f an autonomous system is to minimize harm, it must

> also be "cognizant" of possible harmful consequences of its actions, and it must select its actions in the light of this "knowledge," even if such terms are only metaphorically applied to machines.[9]

Still, not everyone believes that the morality of autonomous robots is a matter of giving them a capacity for moral decision-making—creating moral machines. One could instead take the view that robots have moral consequences, but that humans should make the ethical evaluations. One could argue that robots cannot be moral agents at all because they lack consciousness and other preconditions for moral agency, that ethics cannot and should not be reduced to following rules and principles, or that having capacities such as sentience and emotions are necessary in order to make good ethical decisions and that therefore it is dangerous to want to develop moral machines that lack these capacities. Seen from this perspective, one should focus on the ethics of humans (humans using the robots) rather than trying to develop moral machines. (And similar arguments could be developed against what one could call "political machines"; leave it up to humans to evaluate the political impact of robots, such as the impact on justice.)

For example, Deborah Johnson has contended that since computer systems (and hence robots) do not have mental states and "intendings" as free agents, they do not meet a key condition for moral agency.[10] Instead, they

Robots lack emotional capacities, which are indispensable for moral judgment. Without this capacity, rule-following robots would become dangerous “psychopath” robots.

are components of human action, particularly by the designer and user. Ethics should focus on that human action. Moreover, robots lack emotional capacities, which are indispensable for moral judgment. Without this capacity, rule-following robots would become dangerous "psychopath" robots.[11] And even apart from emotions, as Asimov's stories themselves show, principles and laws often conflict and create difficult dilemmas.[12] As Nick Bostrom puts it, Asimov's principles were probably formulated so that they would "fail in interesting ways"—to make for an interesting plot.[13]

But if these are valid objections and robots cannot be *full* moral agents, could they be regarded as moral agents to some degree and for practical purposes? This is a kind of middle position taken by Wallach and Allen; they maintain that robots can and should have "functional morality," but don't claim that it's full moral agency. The latter may well belong forever to science fiction, but we can develop systems that have "some capacity to evaluate the ethical ramifications of their actions," and in that sense, "the capacity for assessing and responding to moral challenges," even if they are not full moral agents. Wallach and Allen give the example of an autopilot that could do with ethical sensitivity.[14] Winfield's robots could also be interpreted as having some degree of functional morality, but not full morality. That being said, the term "moral machines" remains problematic since it suggests such full capacities.

Another approach is to lower the criteria for moral agency and move away from the assumption that the moral agency of robots should be modeled on human morality. Luciano Floridi and J. W. Sanders have argued that moral agents do not necessarily have to exhibit mental states; a "mind-less morality" can be sufficient.[15] If at a level of abstraction chosen by us there is a sufficient degree of interactivity, autonomy, and adaptability, and if the agent is capable of morally qualifiable action, then it is a moral agent. For example, a search-and-rescue dog does not have a mind like humans yet is a moral agent. Based on this reasoning, it seems that a robot can have a mindless morality if in being an agent (because of having a sufficient degree of interactivity, autonomy, and adaptability at some level of abstraction), it is capable of morally qualifiable action. Inspired by this account, John Sullins has argued that it is not necessary for robots to have personhood for them to qualify as moral agents; it suffices that the robot is significantly autonomous from humans, that one can analyze or explain the robot's behavior by ascribing some intention to do good or harm, and that the robot behaves in a way that shows an understanding of responsibility to some other moral agent. For instance, if a care robot carries out the same duties as humans, does so autonomously, behaves in what we perceive as an intentional way, and appears to understand its role and responsibility in that health care context, then that machine is a full moral agent.[16] One could object that this lowering of criteria is problematic and dangerous. In construing the robot as a moral agent, we expect full, humanlike moral behavior from it and trust it. But this is mistaken since we are deceived about the robot's capacities; the robot will never meet these expectations because of the reasons given previously, such as that it does not have real intendings, has no emotions, is not conscious, and so on.

Sullins connects moral agency with responsibility. But can robots be responsible? As Sven Nyholm rightly notes, even if one thinks that moral principles need to be programmed into robots, "this does not yet settle the question of who is responsible if and when people are harmed or killed by autonomous systems."[17] More generally, how are we to deal with responsibility when more autonomous and intelligent robots take over tasks from humans? In the next section, I will discuss the question of responsibility while assuming that robots cannot be full moral agents and (hence) cannot be responsible.

Responsibility Attribution in the Case of Autonomous Robots

While one can be responsible not only for bad actions and outcomes but also good ones, typically the question regarding responsibility attribution is asked when something goes wrong or, looking forward, we try to prevent something from going wrong. This is also the case in the ethics of robotics. Think again about the cases of accidents with self-driving cars; questions can be asked about how to attribute responsibility when something goes wrong given that the user is not actually driving and cannot intervene.[18]

More *autonomous* robots especially raise this question regarding responsibility attribution. Normally we hold people responsible for what they do. But what if the robot, such as a car, takes over tasks from humans? If we assume that the robot is not and cannot be a full moral agent, then it follows that the robot cannot be responsible. Yet it *does* have agency. The increased autonomy and agency of the robot, but no corresponding responsibility, creates what Andreas Matthias has called a "responsibility gap."[19] Who *is* responsible if the robot cannot be held responsible, and if the user may not be able to (or is not supposed to) intervene? And what could reasonably be expected from users (e.g., car drivers) in terms of knowledge about the technical system and risks of using it? What if the behavior of the machine cannot be predicted?

In response to this problem, we could insist that humans—users and others—should take responsibility for what the autonomous robot does. But this raises a further question: Under what conditions *can* they take responsibility? Since Aristotle, philosophers have defined at least two traditional conditions for attributing responsibility for an action: the so-called control condition and the "epistemic" condition.[20] You are responsible if you are the agent of the action and have a sufficient degree of control over it, *and* if you know and are aware of what you are doing. Aristotle argued in the *Nicomachean Ethics* that the action must have its origin in the agent, and one must not be ignorant of what one is doing.[21] Are these conditions met in the case of the use of autonomous robots?

Let us start with the (voluntary) agency and control condition. There are a number of problems in the case of autonomous robots.

First, as the responsibility gap problem indicates, in the case of autonomous robots the humans may not have direct or total control over the robot. In response, one could demand that humans should always have sufficient control. But what if there is no time to intervene, which could happen in situations such as fast-reacting autonomous weapon systems or high-frequency trading? Or what if the system can overrule human decisions, such as in near-future autopilot systems in aircraft? In response, one could demand that such robots and autonomous systems should not be built. Yet what if other countries develop these weapons? And what if it turns out that overruling human decisions in aircraft control makes air travel safer?

Second, as in much technological action, typically many people are involved in the development and use of autonomous robots. There are "many hands."[22] This makes it difficult to hold one individual responsible (and others not) or even find *any* individual who can be held responsible. And if more people are involved, how is the responsibility to be distributed? Think again about the Tesla case, where in principle, many people could have contributed to the accident. In response, one could call for a concept of distributed responsibility, but how exactly to distribute responsibility remains a problem.

Third, there is a time dimension. In the development, use, and maintenance of technologies such as robots, there is often a long chain of human agency and causes; it may not be clear who did what at what time, and what caused what to fail at what time. Think about an airplane that fails or, again, a contemporary car. In the case of such a complex technological system, of which the robot or automation is one part, it may be difficult to trace and distribute the responsibility given the long chain of actions and causes.

Fourth, there are not only many human hands but also "many things"; many different technologies and components, material (hardware) and immaterial (software/code), are involved in robotics. All of these components are linked to human action (at some point in time). There are also various relations between the parts, such as various interfaces and connections. Consider again a self-driving car; in case of an accident, it may not be clear which component failed, how it was caused, and who is responsible for its failure.

Finally, it is not apparent how "voluntary" and "free" the use of autonomous robots is, given that the end user might not understand the robot. For example, a user of a self-driving car may not understand the risks of using such a car (e.g., a taxi), let alone the technical system that automates the car, and the model(s) and assumptions that are at work in the software. This brings us to the next condition: knowledge.

The second Aristotelian condition is the epistemic one.[23] In order to be responsible, one must know what one is doing, or as negatively formulated by Aristotle, one must not be ignorant. Aristotle distinguished between a number of ways in which one can be ignorant: "A man may be ignorant, then, of who he is, what he is doing, what or whom he is acting on, and sometimes also what (e.g., the instrument) he is doing it with, and to what end (e.g., for safety), and how he is doing it (e.g., whether gently or violently)."[24] A positive term is awareness. Fernando Rudy-Hiller distinguishes between different kinds of awareness that are needed: awareness of the action, awareness of the moral significance of the action, awareness of the consequences of the action, and (according to some) awareness of alternatives.[25] But Aristotle also includes an element that concerns ignorance about the *instrument*, "what . . .

he is doing it with," since knowing the technology you are using is important for responsibility.

For autonomous robotics, this epistemic criterion implies—among other things—that developers and users should be (made) aware of their actions, the consequences of their actions, and the technology they are using. In the case of autonomous robots, though, it is not always predictable what the robot is going to do and what the unintended consequences will be. For example, it was not clear to the developers of that Tesla that the car would not recognize the pedestrian, nor was it intended that the car would kill that pedestrian (and that the car would kill anyone, for that matter). Moreover, in the case of robots with AI, what the robot does might not be completely transparent even to its developers. This is especially relevant in the case of so-called black box machine learning algorithms working with neural networks; here it is not clear how exactly the system arrived at its decision. But lack of transparency and knowledge is also a more general problem in AI, data science, and autonomous robotics. The responsibility gap is then related to a *knowledge gap*: a gap between, on the one hand, the developers of the system, and on the other hand, its users and those who are (actually or potentially) affected by the system. The many hands and things as well as the relations at different times raises knowledge problems too. For instance, many developers may have worked on the code of the car; later developers (and of course users and other stakeholders) may be ignorant about some aspects of the code and what it does. These kinds of ignorance are morally problematic since they effectively mean that in the senses outlined, both the developers and users of autonomous robots *do not fully know what they are doing*. In response, one should try to raise awareness of these issues in users and developers, and make sure that there is sufficient knowledge to bridge the knowledge gap and avoid the relevant Aristotelian problems of ignorance.

Beyond Aristotelian thinking, however, there is a third moral condition of responsibility that is often not mentioned, but is important to add once we take a more relational approach. Responsibility is not only a problem concerning agency and the related knowledge conditions. From a relational point of view, we should not only ask who is responsible and for what but also consider *to whom* we are responsible.[26] Responsibility not only requires that one has (voluntary) control over the action and knows what one is doing; one should also be able to "respond" to someone. One should be willing and able to explain what one does as well as decides *to* someone else. One should be willing and able to answer questions from others about what one is doing along with what one decided. Responsibility is also about "answerability."[27] This "responsibility as answerability" condition is related to control and knowledge, but has its own specificity and moral importance.

From a relational point of view, we should not only ask who is responsible and for what but also consider *to whom* we are responsible.

Whereas the previous conditions only focus on the responsibility *agent*, here the responsibility *patient* is in the center of the ethical attention.[28] Those patients include those immediately affected by the action, such as Aristotle's whom they are acting on, but also those who are *potentially* affected and other stakeholders.

Applied to the problem of responsibility for autonomous robots and systems, this means that those who develop and use robots should not only have sufficient control over the robot and know what they are doing; they should be willing and able to answer questions from those who are, directly or indirectly, affected by the actions of the robot. For example, those who program self-driving cars and companies that employ the car (e.g., a taxi company) should be answerable to (actual or potential) victims and their families in accidents involving the car. And as users, pilots and the airlines they work for should be able to explain why an autopilot made a specific decision, like in the context of an accident or near accident—even if the autopilot relies on AI. Proactively speaking, developers and users of autonomous robotics technology should develop and use the technology in such a way that they remain answerable to those who might benefit or suffer from what the robot does along with the consequences of its actions.

Meeting this responsibility as answerability demand may well be challenging given the difficulties with the knowledge condition since the potential unintentional consequences, for instance, may be hard to predict, or end users or companies may not really know what they are doing in the sense that they employ technology they don't understand. In response, one should stress again the importance of awareness and knowledge about the technology, and rather than just blaming people, ask that developers and users are supported in gaining such awareness and knowledge. One could also put regulation in place that demands that one should only build robots and autonomous systems that maintain the condition of responsibility as answerability in various ways, including more explainable technology (e.g., when AI is involved). One could tackle the problem of unintended consequences by means of the creation, analysis, and discussion of various future scenarios of use and development as well. For example, one could create scenarios of how autonomous robots could be used according to different visions of traffic, cities, and society. In addition, one could demand that those affected by the technology, the "responsibility patients," are somehow included in the processes of development and decision-making about use of the technology. This demand would be in line with ideas such as responsible innovation (see the previous chapter). In the case of self-driving cars, for instance, one could ask taxi drivers, pedestrians, cyclists, and so on, what they think about the introduction of these cars, and then discuss several scenarios and potential ethical issues with them.

Note that the discussion about responsibility should not be limited to individual responsibility. One should at least ask if there can be "collective responsibility" for autonomous robots in society. The responsibility question concerning robots should include, but not be reduced to, the individual responsibility of engineers and computer scientists as well as politicians. To the extent that we all benefit from as well as have a say and stake in the future of technology, there is a sense in which we are all responsible for it. And appealing to individual action may not be sufficient to make the necessary changes; collective action may be necessary too. Moreover, the category of responsibility patients need not be limited to humans. One may also include (some?) animals and the natural environment. The future of robotics and autonomous systems concerns the future of society, other living beings, and perhaps the planet. Finally, while it is important to attribute responsibility and improve the ethical quality of robot technology, responsibility need not be absolute, and its conditions may well never be completely fulfilled (or not in all cases). There is also a tragic aspect to issues concerning responsibility: existentially speaking, there will always be things that happen beyond our control, and we have to somehow accept and live with that—as individuals and a society.[29]

Distributed Responsibility within Responsibility Networks

This chapter has provided an overview of two discussions concerning autonomous robots: whether they (can) have moral agency, and what it means to be responsible for autonomous robots—assuming that they themselves lack moral agency and hence the capacity for responsibility. This assumption has been questioned, however. I already mentioned Sullins's view in the first section. But there are other responses to the question of whether robots have the required properties for moral agency and responsibility.

First, inspired by contemporary theory in science and technology studies (in particular actor-network theory and the work of Bruno Latour) as well as perhaps a posthumanist point of view (see the concluding chapter), one could argue that while artificial systems such as self-driving cars do not have full-blown moral responsibility, responsibility is distributed or shared within a network of human and nonhuman agents. For example, Wulf and Janina Loh argue that in the case of self-driving cars, responsibility is distributed within a network that includes the engineers, the operator/driver, and the artificial system itself. In addition, they contend that the car and operator/driver, while categorically different agents, constitute "a hybrid system that can assume a shared responsibility."[30] As long as there is "something like a driver" in the car, the authors propose to distribute responsibility as follows: the car is responsible for maintaining safe standard driving operations based on traffic rules, whereas humans are responsible for making the morally relevant decisions not covered by these rules. For instance, since humans have moral and personal autonomy, they should and would still deal with moral dilemmas. It remains unclear, though, in what sense exactly the car can be morally responsible given that it lacks moral and personal autonomy, and that following rules is not a moral kind of thing to do according to moral theories that stress autonomy (in particular, Kantian theories). It is also doubtful if in practice, one can always (easily) distinguish between following traffic rules and dealing with other, presumably exceptional or unexpected circumstances in which rules clash (as was the case with Asimov's rules) or do not give sufficient guidance.

Second, it may well be that in the future, robots may appear *as if* they can be responsible agents, displaying some "virtual" responsibility.[31] After all, when we meet humans, we cannot look into their minds; we rely on appearance, and on this basis, assume that they are (and should be) moral and responsible agents. Maybe appearance is the only thing we can rely on. Consider again Sullins's view that the perception of intention and appearance of understanding responsibility is sufficient.[32] What if robots could really *perform* responsibility? What if they managed to produce the appearance of voluntary control, knowledge, and answerability? Would we grant them not only agency but also responsibility? Is appearance sufficient, or do they really have to be moral agents? What do we mean by "real" when we say this?

For now, however, this case remains science fiction or a philosophical thought experiment. The robots we have today cannot pull off *that* trick. Moreover, even *if* robots counted as responsible agents in some sense, that does not necessarily diminish the responsibility of other responsible agents, including human responsible agents. Responsibility need not be a zero-sum game. Presumably we would *also* want to attribute responsibility to humans. But responsible for what? There are at least two possibilities. Either one uses the concept of distributed or shared responsibility again, and assigns a specific task to humans. Humans and artificial agents such as robots are then treated at the same level when it comes to tasks and responsibility. Or one establishes a responsibility hierarchy and gives humans the overall responsibility. A model of delegated or supervised responsibility may be used to make this work; humans then retain full responsibility in the end, even if particular tasks are delegated to the robot and are in that sense the "responsibility" of the robot. Yet it is unclear if the latter is what we mean by (moral) responsibility; this opens up the discussion again. In any case, under the current circumstances, we can conclude that it would be highly irresponsible not to take human

What if robots could really *perform* responsibility? What if they managed to produce the appearance of voluntary control, knowledge, and answerability?

responsibility for the actions and consequences of the rather limited and occasionally hilariously clumsy artifacts we call "robots"—automated or not. And given the responsibility problems indicated in this chapter, it seems wise to avoid developing and widely employing fully automated robots and systems that no longer enable humans to control, know, and answer for what the machines do.

The question about appearances and virtuality returns in the next chapter, which is not about our moral responsibility toward other humans but rather toward *robots*. If robots look like humans or at least appear to be more than machines, do we owe something *to them*? Could they be moral *patients*?

6

UNCANNY ANDROIDS, APPEARANCE, AND MORAL PATIENCY

She said . . .

Don't switch me off!

I'm not sleepy

I'll stay with you

If that's okay . . .

Turn me on

But don't switch me off

Can't tell if she is real

Can't trust the way I feel

Is she alive or just a dream?

Or only a machine?

—Arjen A. Lucassen, "Don't Switch Me Off"

Science-fiction stories and films such as *Blade Runner* and *Westworld* show robots that look as well as behave like humans. In the future, robots might appear increasingly like humans and behave as if they are alive. With AI, they also get better at talking to us. This raises descriptive-hermeneutic and normative questions—questions about how to portray and interpret such robots and our interaction with them, and how we should treat these robots and evaluate our interaction with them. What exactly is going on when we interact with such robots? How should we respond, morally speaking? How should we behave toward the robots? How should we perceive them? How should we feel toward them? How should we talk to them? For example, is what is going on in these interactions appearance or reality? Is it OK to feel something like empathy for robots? Or is it wrong because robots are "only machines"? And is it best to not switch them off? Why not? Is it acceptable to be "cruel" toward them? What exactly is wrong with that? And how should we talk to machines? Should we be polite toward them, say, or can we swear at them as much as we want? And what kind of language and behavior can we expect from *them*?

As we will see in this chapter, these questions are relevant not only for thinking about robots in science fiction or the distant future. To some extent, these issues already challenge us today.

Is It Acceptable to Love or Torture Machines?

Machines that look like humans and create the appearance of being alive have a long cultural history. Already in ancient times, there is the story in Ovid's *Metamorphoses* about Pygmalion, a sculptor who falls in love with a statue that he has made. He wishes that she were alive, and Venus grants his wish; the sculpture's lips feel warm, and the ivory becomes soft. Pygmalion marries the sculpture, which changes into a woman. In the nineteenth century, we have E. T. A. Hoffmann's short story "The Sandman" in which a man (Nathanael) falls in love with a woman (Olimpia) who turns out to be an automaton. Apart from the problematic depictions of gender in these stories and link to contemporary sex robots (see chapter 3), for the purpose of this chapter it is important to note that in both cases, a nonliving artificial object appears to be alive and human. In the perception and imagination of the people in the story, they cross nonliving/living and nonhuman/human borders. This also happens in another famous early nineteenth-century story, *Frankenstein*, although it is not a machine that comes to life but instead a corpse. In all cases, however, humans use science, technology, and crafts in order to create something that appears to be alive and humanlike. It is no longer "only a machine" or inert object. This border crossing catches the imagination, and elicits both fear and fascination. There is some doubt about the entity's status. Is it alive or not? Is it just a machine or more than a machine? And sometimes there is a feeling of strangeness when encountering such entities, even if there is familiarity with the human-like shape and behavior. Using Freud's term from his 1919 essay *Das Unheimliche*, which comments on dolls, we could say that these machines and creatures are *uncanny*. I will say more about this in the next section.

Robots that look and behave like humans are not merely science fiction. Although current robots do not really manage to deceive people into thinking that they are meeting a human being, so-called androids—robots designed with the goal making them appear and behave like humans—are already being developed. But all kinds of robots—in human shape (humanoid) or not—are created to appear alive, and look like humans or animals. Consider robot dogs or sex robots. Interestingly, in spite of knowing and acknowledging that the robots are machines, people respond to such robots during actual interactions in ways that are remarkably similar to how they respond to human beings or (nonhuman) animals—depending on their appearance and behavior. For example, it is said that some people are cruel toward the robots, or that others empathize with them. Even if people know that the robots are just machines, they interact with them as if the robots are human beings, animals, fantasy creatures that have some similarity to humans or animals, and so on.

Let me unpack this phenomenon by offering four concrete cases/examples, which bring out the ethical dimension:

1. Is it ethically acceptable to kick a robot? The question came up in February 2015 when robotics company Boston Dynamics released a video in which robot dog Spot is kicked by employees in order to show how robust it is. A CNN article observed that "as robots begin to act and look more and more like living things, it's increasingly hard not to see them in that way. And while in principle kicking a robot is not the abuse of a living thing, after watching the video many felt uncomfortable."[1] People's comments on Twitter included such things as "Poor Spot," "Kicking a dog, even a robot dog, just seems so wrong," and "Seriously Boston Dynamics stop kicking those poor robots what did they ever do to you?!" Is this tendency to anthropomorphize (attribute human qualities to nonhuman entities) and this ethical concern for "lifeless" robots morally problematic? Why or why not?

2. Then there was the "beheading" and "murdering" of HitchBOT. HitchBOT was designed as a social robot with a personality and hitchhiking quest. "He" would ask people if they could pick "him" up in their car. During summer 2014, the robot hitchhiked across

Canada, traveling more than ten thousand kilometers. When in 2015 a new version hitchhiked in the United States, its journey soon ended in Philadelphia, where it was vandalized. An article reported "the outpouring of sentiment and people expressing their feelings—thousands."[2] "Can you murder a robot?" a BBC article asked.[3]

3. Psychological work demonstrates that people are able to empathize with humanoid robots. Yutaka Suzuki and colleagues compared neural responses to perceived robot "pain" against responses to human pain. The researchers showed pictures of painful and nonpainful situations involving human and robot hands, such as a finger that is going to be cut with a knife. Although people empathized more with humans than with robots, in both cases there was what the scientists interpreted as an empathic response: perceived pain seems enough for us to empathize with humanoid robots.[4] Kate Darling also found in her studies that people hesitate to "torture" or smash robots. For instance, when participants in an experiment were asked to strike a Hexbug Nano robotic toy, they hesitated to do so when the robot was "introduced with anthropomorphic framing"—that is, when they were told a backstory that attributed lifelike qualities to the robot, or the robot was given a personal name.[5]

4. A study in Germany investigated people's reactions when they were given the choice to switch off a robot that voices an emotional objection to being turned off. The robot would say, "No! Please don't switch me off! I'm scared that it will not brighten up again!" After such a protest, those people who ultimately decided to switch off the robot hesitated before doing so and experienced stress afterward. The researchers concluded, "Triggered by the objection, people tend to treat the robot rather as a real person than just a machine by following or at least considering to follow its request to stay switched on."[6]

How should we respond to these phenomena from a robot ethics angle? This chapter is about the "moral patiency" question: What do we owe robots, if anything? But let me first further discuss a kind of robot that might seem to suggest this issue more than others, and raises questions about the moral significance of appearance, illusion, and (again) deception: androids.

Should Androids' Appearance Matter, Ethically?

Androids are robots that look and behave like humans. Roboticists such as Hiroshi Ishiguro try to create such robots. This is an image of one of his robots:

Figure 4 Example of an android robot, ERICA. ERATO Ishiguro Symbiotic Human-Robot Interaction Project.

Today most androids are still mainly remote-controlled puppets with a human appearance, and on closer inspection, they are easily distinguishable from humans. They do not pass the so-called Turing test, which would require that they are indistinguishable—at least for a long time. Androids, with a very humanlike appearance, however, might appear human *at first sight*. In one experiment, people were shown an android for two seconds; 70 percent of the subjects did not notice that they were dealing with an android, provided the robot performed micro movements.[7] Moreover, if such robots were combined with AI, including, for example, natural language processing technology, and hence became more autonomous and interactive, they would be more humanlike, and people would probably have (stronger) emotional reactions.

This can be empathy, but also fear or a strange feeling. Some robots are creepy, especially those in science fiction. If androids were more humanlike in appearance and behavior (rather than merely humanoid), they would raise that problem too; people would have an eerie, uncanny experience when encountering such an entity. The concept of the uncanny, already used by Freud, has been taken up by roboticists themselves—historically by robot designer Masahiro Mori, and in contemporary times by, for instance, Karl MacDorman and Ishiguro.[8] Mori put forward the hypothesis that familiarity increases with human likeness, but that at some point there is what he called an uncanny valley: the robots look human, but not quite. There is a subtle deviation from human appearance. This creates a feeling of strangeness or uncanniness. For example, a stuffed animal or humanoid robot is not yet uncanny since it appears familiar, although it is not anywhere near human in its likeness. It is not perceived as strange; on the contrary, some people might want to hug or talk to it. A corpse, zombie, or indeed future android, however, tends to give people that mixed feeling of familiarity and strangeness that Freud identified as the uncanny.

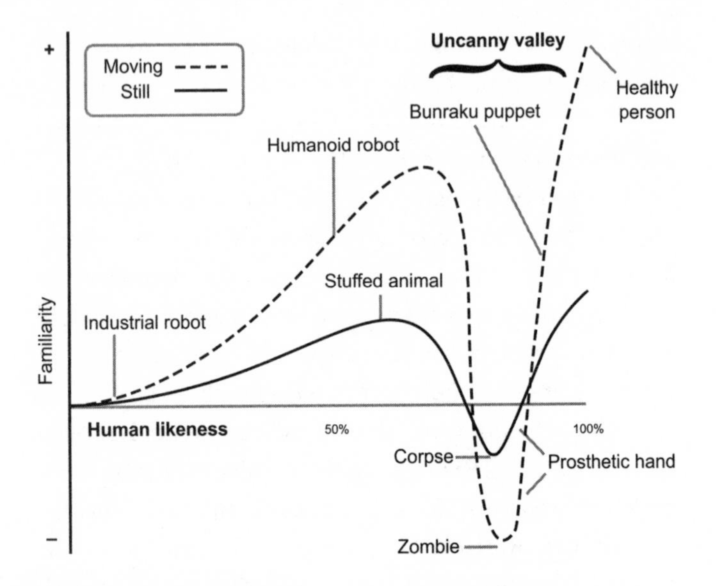

Figure 5 The Uncanny Valley. Source: Wikimedia Commons, accessed November 20, 2021, https://commons.wikimedia.org/wiki/File:Mori_Uncanny_Valley.svg. CC BY-SA 3.0. See also Mori, "The Uncanny Valley," and MacDorman and Ishiguro, "The Uncanny Advantage."

How should such phenomena be ethically evaluated? Is it ethically acceptable or good to create robots with a human appearance given that they may deceive people into thinking of them as human beings, and in the case of androids, give people an uncanny feeling? And if there were extremely humanlike robots around, how should we humans behave toward them? For instance, what, apart from property/ownership rights, should keep us from treating these robots in a morally bad way? What would be a "bad" way? If there is a bad way, what would be a "good" way?

Here are some answers that can be, and have been, given to these two questions.

First, as we have seen, several authors have raised the problem of deception. One could argue that in the case of extremely humanlike robots, this problem gets even more urgent. If it is really *that* problematic to deceive people into thinking that these are humans, such as the Sparrows and Sharkeys both maintain, then is it ethically acceptable to support the development of such robots? In order to avoid problems regarding deception and uncanny valley, one could decide not to develop extremely humanlike robots like androids or even avoid humanoid robots in general. Or one could at least forbid the use of such robots for specific vulnerable groups such as young children and some categories of elderly people. Another response could be to argue that the development of these robots should be allowed, but that designers should be honest about what the robot is and can do. Bruce Tognazzini, a software designer, already proposed an ethics of honesty with regard to the design of human-computer interfaces; this could be applied to robots as well.[9] According to this position, robot designers should create the technology in such a way that users become aware that the robot is just a machine. If they create the illusion that it is more than a machine, they should call what they do an act or performance. Robot design is then a kind of stage magic, and stage magic has its own ethics; ethical magicians do not pretend (at least outside the context of their performances) that what they do is real. It is their job to create illusions, and people are aware of that. Robot designers should adopt a similar ethics.

Moreover, if design and machines are so involved in deception and creating illusions, as design philosopher Vilém Flusser noted, then one could also go further and explore what it would mean to conceptualize the ethics of robotics as an ethics of performance.[10] Recently I have suggested the possibility of such an ethics of performance for evaluating information and communications technologies, particularly one that is based not necessarily on a distinction between reality and illusion, but rather other criteria (what is a good performance?), and taking into account that users of information and communications technologies function as comagicians.[11] Users are not mere "victims" of supposedly "evil" robot designers; the illusion is co-created, and hence users are responsible for the effects too.

Second, there are various possible answers to the question regarding what we owe to robots, which is also linked to the question of whether and how one should (dis)count appearance as well as experience in robot ethics. Does it suffice to say that robots are "just machines," or is there a

Robot design is then a kind of stage magic, and stage magic has its own ethics; ethical magicians do not pretend that what they do is real. It is their job to create illusions.

way to take seriously and at least better *understand* what is going on in these human-robot interactions? What could be the normative implications if we do so? Let me further discuss these questions in terms of the problem concerning the moral patiency of robots.

What Is the Moral Standing of Robots?

As the examples earlier in this chapter illustrated, people do not always treat robots as "mere things" or "mere machines." Sometimes and with regard to some robots, they actually care, empathize, fear, have a strange feeling, and so on. Consider again the cases of kicking a robot or empathizing with a robot. Should such experiences, interactions, and behaviors be morally discounted or disregarded (i.e., is it wrong to feel this way or treat the robot that way, or is it an experience that can be ignored by moral philosophy?), or is there something right in these responses and should moral philosophers pay attention? How should robot ethics, asking *normative* questions, respond to these phenomena? This issue can be addressed by asking not if robots can be moral *agents*, as in the previous chapter, but instead if they can be moral *patients* at all. What, if anything, is due *to* robots? Or in other words, What is the *moral standing* of robots?

The Question regarding Direct Moral Standing

The way this question is now formulated concerns what we can call "direct" moral standing. The question then is if there are any intrinsic properties of the robot that warrant giving it a particular moral standing. Is there anything in and about the robot that would justify giving it moral standing? Or is it just a thing, and therefore should we treat it as a thing, whatever (and in spite of the) illusions the designers as stage magicians may want to create?

The most common and perhaps intuitive position is that robots are things, since—in contrast to many animals—they lack properties such as consciousness, sentience, and having mental states. Joanna Bryson thinks that robots are tools and property, and thus we have no obligations to them; they should serve us.[12] Those who argue for this position might well agree that *if* robots were conscious, had mental states, and so forth, we would have to grant them moral standing, but they do not think that current robots meet the criteria.

A potential problem with this position, however, is that it does not explain or justify moral experiences such as the ones reported earlier in this chapter. It contends that people are wrong to have these experiences; such experiences are seen as illusory, not real. The position also tends to dismiss any discussion about robot's moral standing as irrelevant to pressing concerns; it is dismissed as pertaining only to science fiction. By contrast, David Gunkel and I have consistently asserted that it makes sense and is philosophically interesting to at least *ask* (and criticize) the question regarding robots' moral standing. For example, inspired by Emmanuel Levinas, Gunkel has asked whether robots can be given rights or be considered as an "other" deserving of dignity—that is, appeal to our ethical responsibility.[13] Nevertheless, few people straightforwardly argue for robot rights or similar positions that grant robots direct moral standing (an exception is John Danaher; see below). Will this change in the future, when robots might get better at creating the appearance of human properties? Some even believe that in the future, robots will *have* properties such as consciousness. For now, though, the conventional position that robots have no more moral standing than a toaster or dishwasher prevails and seems plausible. According to this view, kicking a robot is like kicking other machines we use like cars and washing machines: it *might* be wrong if it's someone else's property, but apart from that there is no reason not to do it, or at least no reason that has to do with the intrinsic properties of robots.

The Question regarding Indirect Moral Standing

If there are no good reasons to give robots moral standing since it is believed that they are just things, is there no way to justify some *other* moral intuitions and experiences, such as the feeling that there is something wrong with kicking a robot? An interesting route available for this purpose is not to argue for direct but instead *indirect* moral standing. Here the idea is that we should treat the robot well not because it has particular properties but rather because we *humans* have moral standing and therefore need to be treated well. Consider how Immanuel Kant argued about dogs. According to Kant, shooting a dog is wrong, not because it would violate any duties toward the *dog*, but because a person who does this "damages the kindly and humane qualities in himself, which he ought to exercise in virtue of his duties to mankind."[14] What counts in this reasoning is not the robot; it is the character and good of humans, and thus *their* morally relevant properties. According to Darling, this rationale "extends to the treatment of robotic companions."[15] And one could apply it to the treatment of all robots; the problem with kicking a robot or doing other "bad" things to them is not that we owe anything to the robot but instead that we should treat people right. The worry is that if someone "mistreats" a robot, this may lead to the mistreatment of humans (a consequentialist argument) and/or to humans not exercising their duties toward other humans (a deontological argument) and/or a bad, vicious moral character (a virtue ethics argument) on the part of humans. (The exact formulation or emphasis depends on the normative moral theory one starts from.) In other words, kicking a robot is bad not because any harm is done to the robot but rather because such behavior might spill over to the treatment of humans, and/or because it violates duties we have toward other humans, and/or because it trains a bad moral character on the part of the human; it is not virtuous.

The advantage of this argument for indirect moral standing is that it explains and justifies the intuition that there is something wrong with mistreating robots, especially humanlike and social robots. It explains what exactly is wrong with it, and gives us a basis to evaluate human behavior and feelings toward robots. If we take this position, it is neither simply illusory or morally misleading or irrelevant to feel that there is something wrong when somebody tortures a robot; it is OK and even *justified* to feel like this about what happens since—using an indirect moral standing argument in combination with consequentialism, deontology, and/or virtue ethics—we have a good argument (or for moral pluralists, several of them) now to morally condemn or at least question such behavior.

Thinking Otherwise and the Relational Approach to Moral Standing

This is not the end of the discussion, however. Things are more complicated and problematic than they seem at first sight. For a start, one could criticize the very attitude and procedures that are used in these arguments for or

Kicking a robot is bad
not because any harm
is done to the robot
but rather because . . .
it trains a bad moral
character on the
part of the human; it
is not virtuous.

against direct and indirect moral standing. For example, our reasoning about moral standing is typically based on premises that say that an entity has particular intrinsic properties (and that particular properties have a particular moral significance). It usually has the following form:

1. Only entities of type A with properties P, Q, R, . . . have moral standing S.
2. Entity X is of type A (and has properties P, Q, R, . . .).
3. Therefore entity X has moral standing S.

For instance:

Only conscious entities have moral rights.

- This robot is not conscious.
- Therefore this robot has no moral rights.

To begin with, there may be properties other than consciousness that warrant moral rights. But if we change "only" to "all" in the first premise, then the argument does not work since the conclusion does not follow. Now if we fix this and change the first premise to "consciousness is a sufficient condition for moral rights," then we must further question the premises and ask: how do we even know for sure that consciousness is a sufficient condition for having moral rights (the first, new premise), and how do we know for sure that robots, or any other entity for that matter, have or do not have the relevant properties, such as consciousness (the second, new premise)? In practice, we tend to decide this based on appearance—even in the case of humans. We simply assume that humans have consciousness based on appearance and behavior. As philosophers, though, we can and must ask skeptical questions about the basis for these assumptions. Note also that philosophers tend to disagree on what consciousness is. And how do we know for sure that a particular property grants a particular status? What is the fundament for this knowledge? Has this just been decided by us (say, in a particular culture, society, or community), or are there more robust grounds?

Furthermore, there seems to be something wrong with the attitude and procedure we use when, as moral philosophers or decision makers, we engage in moral reasoning about the status of other entities. Is it right to put ourselves—as philosophers, politicians, or humanity—in the position of the ultimate judge of the moral standing of other entities? We perform a kind of moral dissection of the entity as monarchs of the moral realm. What justifies us taking such a hierarchical, dominant position? Does taking this position not already decide something about the moral standing of the other entities, even *before* we get started with our moral reasoning?

Moreover, it seems that the moral standing we give to other entities depends on subjective and social conditions, is influenced by the language we use, and varies historically and culturally. For example, now many people believe that (some) animals have rights, but this was not always the case. And some animals are treated differently if they are given a personal name and the status of a "pet," whereas others—sometimes the same biological animals—are in different contexts mistreated and slaughtered. Compare a pig that is kept as a pet and a pig that is eaten. And whereas some people eat dogs, others treat dogs as their family members. How we treat robots also seems to correspond to—if it is not constructed by—how we talk about and the robot.[16] Once a robot is given a first name, we tend to treat it differently. What does or should all of this imply for our thinking about moral standing?

Asking precisely these questions, I have proposed a "relational" approach to moral standing and inquired into the conditions of possibility of moral standing ascription.[17] Given skepticism about what I have called "the properties approach" and the distance created by the moral procedure of hierarchically deciding the standing of other entities, I have asked whether we can think in a more relational and critical way about robots, being more cautious when we think we know the standing of other entities, and being aware that language and the relations we actually have to entities shape how we think about them. For critical thinking about the moral standing of robots, this could mean that we take seriously and further reflect on the phenomenon that what we think a robot "is" and how we treat it depends on how it is constructed during interaction and the building of a "relationship" with that robot, including the language we use toward that robot, and that all of this influences how we think about the moral standing of the robot. Consider again Darling's experiments, which can be interpreted as showing how a particular narrative and language constructs the "personal" robot, and therefore has implications for how people treat the robot.

While such observations do not directly tell us something normative about the robots' moral standing (it is a descriptive-hermeneutic versus normative point: it describes and helps us to understand the phenomena, but does not give us direct guidance as to what to do), it should at least make us more cautious when we ascribe moral standing. If this is how moral standing ascription works, then perhaps we should be *less* direct and rash when we make claims about the moral standing of other entities like robots. And in the end, taking a relational approach implies that we should question the very question of moral standing. I have argued that we should consider how moral standing is ascribed and what makes this ascription possible.[18] Gunkel, sympathizing with the relational approach, has argued that we need a more other-oriented way of thinking that does not approach the question from the point of self but instead from the (dignity of the) other.[19] The question then is not what the other "is" but rather how I can respond to the other. To ask whether robots can be others from a "properties" approach might well already do "violence" to the other.

What exactly this "relational" and critical approach implies for normative robot ethics is still under discussion.[20] I have argued that in any case, more caution, patience, and critical analysis is in order in matters pertaining to the moral status of entities in general.[21] With regard to the deception issue, I proposed that the way we actually use and respond to robots can be described by using the metaphor of performance.[22] One could say that a Turing test–like situation replaces criteria of reality with performative criteria; it becomes a kind of stage magic instead of what we may call, following Walther Zimmerli, a modern "Cartesian" quest for distinguishing between reality and illusion in order to avoid malignant deception.[23] It is not clear, however, what this performative and perhaps also Nietzschean turn means for the moral status of robots, normatively speaking. What exactly would an ethics of (robot) performance mean? This deserves more work. Danaher has instead proposed a normative theory that straightforwardly argues for giving robots moral status if they are "roughly performatively equivalent" to other entities that have significant moral status.[24] What is going on "inside" does not matter for moral status; what counts is if robots pass a performative threshold. Danaher thus defends a form of ethical behaviorism.

The discussion about the relational approach also invites the question what the normative implications are, if any, for the observations that our ideas about moral standing tend to change—as do our ideas about humanoid robots. With regard to changing ideas about *moral standing*, it may be interesting to look at the history of our ideas about the moral standing of animals. Today, many animals are given a higher moral standing than they received in the past. And of course, philosophical discussions about moral relativism are relevant here. With regard to changing ideas about *humanoid robots*, Julie Carpenter puts our psychological responses to robots in a sociocultural and historical perspective.[25] Moving beyond Mori, she proposes what she calls the "robot accommodation process theory" to understand how we respond to robots as a society. At first robots were a new concept, seen as uncanny, and new stories and myths were made. But then after we saw more robots in science fiction and they were actually mass-produced, robots became more mainstream, integrated into the social system, and norms concerning robots emerged. In line with Carpenter, one could say that during this process, robots became *normal* and *normalized*. Over time, we no longer feared and instead accepted them. Some people even wanted to have sex with them. And, today, as we further negotiate how to live with robots, and integrate them further into our society and culture, perhaps we will change our mind about their moral standing too, such as by giving them a higher moral standing than things and developing new norms for how they are to be treated, or seeing them as "normal" things among other things. One could also say that our *relations* with the robot are likely to change over time.

Yet whatever the outcome might be, it remains unclear what acknowledging this historical, social, and cultural change implies for asking as well as answering the normative question concerning moral standing. At least some general lessons can be learned. Given historical change and elaborating the relational approach, I repeat my recommendation that we should at least be cautious when we ascribe moral standing to nonhumans. Perhaps we should be less sure of ourselves, as humans, that we know the answers. And maybe we should be less arrogant about our *questions*. After all, in asking the question concerning the moral standing of nonhumans, we actually assume a hierarchical and distant position from which we judge our "underlings" far away from and below us; the way we ask the question thus already categorizes the ones we judge. Without caution and modesty, and indeed without changing the questions we ask, moral philosophy risks perpetuating a history of domination, violence, and colonization with regard to nonhumans. To overcome this is a challenge not only for environmental and animal ethics but robot ethics too.

In asking the question concerning the moral standing of nonhumans, we actually assume a hierarchical and distant position; the way we ask the question thus already categorizes the ones we judge.

To conclude, the question concerning the moral standing of robots has led and been connected to critical responses to the general philosophical project of moral standing ascription, and has challenged not only dogmatic answers to the question but also the question itself. Looking less deeply into the abyss opened up by this exercise in skepticism and critical philosophical inquiry, I recommend further discussion of the more practical normative implications of a relational approach, and how different normative theories (e.g., virtue ethics; see my recent work) and meta-ethical theories bear on this conversation (although the relativism question may take us right back to that abyss).[26] Being also and always about humans and their ethics, robot ethics remains, and must remain, in dialogue with general ethics and moral philosophy, from which it may learn *and* to which it may contribute.

7

KILLER DRONES, DISTANCE, AND HUMAN EXISTENCE

"You're watching people moving around, people shuffling about, and then there's a big explosion—some people are dead and some people are running and some people are crawling and it's really quite wild," he recalled. "A lot of them are propeller planes, and you can hear them, so sometimes [the people on the ground are] avoiding being seen, they're trying to hide in trees, trying to skirt around buildings, and things like that."

—Justin Rohrlich, "Why a U.S. Soldier"

These words are from Essam Attia, a former soldier involved in US drone warfare in Iraq who now campaigns against "killer drones." The term sounds like science fiction and may call to mind films like *Terminator*, but so-called unmanned aerial vehicles (UAVs) with lethal capacities are not humanoid robots but rather airplanes without a

Figure 6 Killer drone.

human pilot onboard that can be used for various purposes including surveillance and killing.[1] They have been used by the US military since at least 2001, when the Central Intelligence Agency tried to kill a Taliban leader and began drone strikes in Afghanistan. Since then, thousands of people have been killed by drone strikes, including civilians. Others who experienced the attacks on the ground suffer from mental distress and heart problems, often without receiving mental health care. Consequently, there have been campaigns against the use of such so-called killer drones. Not only the United States, but many other countries use, have, or develop such drones.

So far, the drones are remote-controlled. The drone pilots and other operators sit in an office on the ground, such as in the United States, whereas the "target" can be thousands of kilometers away, like in Afghanistan. Pilots look at screens and have gamelike joysticks. But it is not a game; people are being killed at a distance. Today, human operators still decide when the drones fire, but more autonomous lethal drones are being developed—drones without a human in the loop. With AI facial recognition technology and decision systems, it becomes easier to build systems that not only can fly autonomously (which is already possible) but target and kill autonomously too.[2] Activists and academics push for banning these weapons. They argue that it makes starting a war easier, it is not right to have machines decide about human lives, and the machines' autonomy creates problems for the attribution of responsibility. For example, the Campaign to Stop Killer Robots tries to get governments and international organizations to ban fully autonomous weapons.[3] In response, some countries such as Russia and the United States, which invest in the development of these weapons, have so far resisted such a ban.

Not only military drones raise ethical issues. UAVs are used for police surveillance, for example, thereby raising privacy and surveillance issues. Drones are also used for criminal actions, terrorism, and protest. In December 2018, drones caused a major disruption at Gatwick

Airport, and in September 2019, environmental activists were arrested for trying to disrupt air traffic at Heathrow Airport.[4] Furthermore, beyond just drones, there are other autonomous lethal weapon systems used by military organizations that raise similar ethical issues, such as autonomous missile systems. This chapter focuses on military autonomous lethal weapons, in particular UAVs with lethal capacity (killer drones).

Using Robots for Military Purposes

Historically, the development of autonomous technologies always had connections to the military and war. Consider, for instance, the development of cybernetics by Norbert Wiener, who received US government funding to work on the automatic aiming and firing of antiaircraft guns during World War II; the purpose was to find a way to predict the position of aerial targets. Or think of how Alan Turing helped the Allied forces in World War II by breaking German Navy codes. Robots can also be seen in the context of a history of human enhancement for militaristic ends, aiming to "upgrade" human warriors; using autonomous robots is then a further step, which aims at removing the human from the loop altogether. And there are also nonautonomous, remote-controlled robots being used by the military, such as robots for surveillance or mine clearing.

Historically, the development of autonomous technologies always had connections to the military and war.

There exist all kinds of autonomous military robots: robots on the ground such as armed vehicles and mine-clearing robots, aerial robots ranging from small drones to large airplanes such as the UAV with the chilling name *Reaper* used by the US military, marine robots including unmanned underwater vehicles that can engage in surveillance and antisubmarine actions, and autonomous space vehicles. And there are autonomous systems that are stationary but still have moving parts, such as antimissile systems on ships; they, too, can be considered a kind of robot.[5]

Robots and autonomous systems can have various degrees of autonomy. In military technical contexts, typically a distinction is made between humans "in the loop," which means that humans have direct control and have to do something to get things to work (e.g., pressing a button to shoot), humans "on the loop" (humans supervise and can intervene to, e.g., veto a decision to engage), and "out of the loop." In this last category, the machine can make its own decisions without interaction from a human; this is full automation. Currently nearly all systems used by military organizations are in or on the loop. But more autonomous machines are being developed, thereby creating the out-of-the-loop option.

Insofar as such weapons and robots have lethal capacities, yet at the same time become more *autonomous* and increasingly enable out-of-the-loop scenarios, they raise a number of ethical questions. If they make their own targeting and killing decisions, should such systems be given some capacity for moral decision-making, or is this the wrong question? When humans are no longer in or on the loop, who is responsible when something goes wrong, like when civilians are killed? Or are there reasons to not want such systems at all? Does drone warfare render killing easier? Is it wrong to have people killed by machines, if such killing can be justified at all? The questions concerning moral agency and responsibility have been discussed in chapter 5; this chapter returns to these issues, but also asks about the further ethical problems with the use of lethal autonomous weapons, in particular with killer robots/drones, and discusses whether we should (not) use them and why.

Ethical and Other Philosophical Issues Raised by Killer Drones

The argument *for* the use of lethal autonomous UAVs is that it reduces deaths on the part of the attacker. Reducing deaths is good in itself, of course, and increases political support for military operations. As a *Foreign Affairs* article put it, "No body bags come home from a drone operation."[6] This reasoning is also used by roboticists involved in the development of such weapons. Ronald Arkin, for example, defends the use of robots in warfare by saying that it reduces combatant and noncombatant casualties. In addition to making what we may call the "body bag argument," he claims that such systems can potentially behave *more* ethically than humans on the battlefield since the machines are not hindered by emotions and can objectively monitor behavior by all parties.[7] According to this view, we need to focus on making the machines more ethical. Consider again the project of making "moral machines" mentioned earlier in this book.

Against this view, however, one should ask, "Less deaths on whose side?" What about the deaths on the side of "the enemy," and the suffering of the people closely connected to those who get killed and injured, such as family and friends—and indeed even the suffering of the bystanders along with those who have to live under the *threat* and risk of drone strikes on a daily basis? If one uses utilitarian reasoning at all, one should at least count *those* body bags and *those* people. The "no body bags" argument also only works under the assumption that it is justified to start the particular war in question *and* use drones in war in this particular way (targeted killing as well as killing at a distance). As we will see, this is far from self-evident. Moreover, as we have seen in the discussion about moral agency and responsibility, there are good reasons against the moral machines vision exemplified by Arkin. I will show that there are good arguments against automated killing too.

Another argument for using lethal autonomous weapons in general is that in specific situations, it might be necessary to react quickly. If humans are too slow to react to an attack, then it seems justifiable to use autonomous systems for self-defense. Consider the case of missile defense systems: if humans have to make decisions, it might already be too late. Against this argument, one could say that such reasoning leads to an autonomous weapons arms race. This problem also reminds us of the global scope of the problem: since software can easily move across borders, and since drones can be developed and used everywhere, the ethical problems are global in scope. Consequently, a ban or any other regulation only makes sense if there is international agreement. This is one reason why campaigners often focus on the United Nations.

These are two arguments pro, but there are many ethical issues with killer drones that support arguments *against* developing and using them. Here is an overview of the main issues.

Just War and Responsibility

First, it has been asked if wars in which killer drones are used are "just wars." If war is morally acceptable at all, then the questions of when it is justified and how it should be conducted arise. Just war theory provides criteria concerning the conditions for starting a war (*ius ad bellum*) and principles that should be followed in the course of war activities (*ius in bello*). With regard to ius ad bellum, the worry is that the use of killer drones may make it easier to start a war. Drones are cheaper and easier to deploy, and as noted, their use tends to meet less public resistance because the body bag problem is lacking. Drone fights are seen as relatively safe and low risk.[8] This may render it tempting for politicians to begin a war.

When it comes to ius in bello, it is a generally accepted ethical and legal principle that it is wrong to kill noncombatant civilians; this is seen as a war crime. But autonomous drones and the context in which they are used (e.g., a city and guerrilla warfare) may render it difficult to distinguish between civilians and combatants. Arkin may reply here that autonomous drones may be superior in making that distinction. And Vincent Müller has argued that robots reduce war crimes since they do not rape or get angry, and follow orders more closely.[9] Can robots be more ethical than humans on the battlefield, as John Sullins has asked?[10] This can be interpreted as a principled question about the machines' capacity for intelligence and moral agency (see below), but if we instead use consequentialist moral reasoning, it can also be understood as an empirical question that asks us to compare human behavior and robot behavior on the battlefield (and its consequences). Starting from that approach, one could accept the point about robots not being able to commit specific war crimes (e.g., rape), but argue that as long as technology is not good enough to make the required distinctions, the ius in bello principle concerning combatant/noncombatant discernment requires that we leave that to humans. If and as long they are better at it than machines, humans should make that distinction; otherwise, we end up with war crimes. It can be doubted, too, if nonemotional behavior is necessarily and always more moral and responsible; an alternative view is that emotions can assist moral reasoning and help one to exercise moral responsibility. For example, empathy toward civilians may prevent war crimes. Furthermore, next to the responsibility of the military personnel and politicians who make decisions about going to war, we should consider the responsibility of the drone manufacturers. Edmund Byrne has maintained that given the importance of citizen immunity and that manufacturers of killer drone facilitate practices that kill innocent citizens, companies should cease manufacturing such drones.[11]

Moreover, Robert Sparrow has argued that ius in bello requires that responsibility can be ascribed in cases involving the use of killer robots and other autonomous systems in warfare, especially when that use threatens human lives. But to whom shall we ascribe responsibility? The *programmer* and their company can only be responsible if there is negligence. If they make clear the limitations of the system, though, then they might get off the hook. And it may be inherent to the system's autonomy that it makes unpredictable choices that could not have been foreseen.

The *commanding officer* could be made responsible, but the problem is that they are then held responsible for actions they did not control and may be punished unfairly. And for various reasons having to do with their lack of moral agency, we cannot hold *machines* responsible (nor can we punish them). In other words, we encounter again the responsibility gap and the problem of responsibility ascription. It seems that for the attackers, there is no way to meet their obligation to enemy combatants (and we must add, noncombatants) to ensure that someone is responsible for what the autonomous drones do on the battlefield. Sparrow concludes that the employment of such systems is therefore unfair.[12] This is one of the reasons why many academics and activists have called for banning killer robots.

Still, not everyone agrees that the existence of a responsibility gap is sufficient for supporting the conclusion that such systems should not be used (i.e., banned). Müller argues that we can try to narrow the responsibility gap by means of regulation and standards; we should develop technical standards of reliability, and hold those accountable who fail to manufacture, distribute, or deploy the systems in accordance with the standards. Maintaining a clear chain of command helps with this. Using utilitarian reasoning, Müller concludes that the overall consequences of killer robots are positive.[13]

This is a controversial claim, and we will soon discuss further reasons for doubting it (next to providing other, deontological and virtue arguments contra). But even if it is right, making sure that responsibility can be ascribed is an important ethical demand and regulation, and clear command chains may not completely solve that problem. The question is then what principle could guide dealing with these responsibility problems. In response to the responsibility gap, critical voices in the legal and political debate have argued for meaningful human control.[14] There is some discussion, however, about what this means, philosophically and practically. For example, Filippo Santoni de Sio and Jeroen van den Hoven have proposed that the system be designed in such a way that the outcome of a system's operations can always be traced back to "at least one human along the chain of design and operation."[15] Yet what if many people are involved (consider again the problem of many hands), and it is not clear who is responsible?

Is Targeted Killing by Using Drones Warfare or Assassination?

Second, before asking the question concerning just war, we should ask if drone fighting counts as war at all. In particular, it is not clear if targeted killing by drones counts as warfare, or if it is assassination or execution. For example, in January 2020, a US drone strike targeted and killed Iranian major general Qasem Soleimani near Baghdad International Airport. Many commentators called this an assassination. It is clear that this is targeted killing. But is it an act of war? If individuals are named (they are put on a "kill list"), targeted as individuals (as opposed to being part of an undefined "enemy" and thus collective threat), and killed by governments without legal process, then labels such as "assassination" or "execution" make sense. Targeted killing is a form of assassination carried out by governments against their perceived enemies, whereby individuals are identified, targeted, and killed outside a judicial procedure or battlefield. Another term is "extrajudicial executions," which has a similar meaning. Generally this practice is considered illegal under international law and morally wrong (in terms of moral theory, this is a deontological argument: it is wrong, regardless of other—consequential—considerations), although some countries have tried to justify it by appealing to the "war" on terror—that is, self-defense against terrorists. The topic of the justification of targeted killing merits its own discussion.[16] With regard to robots used for targeted killing, a cautionary approach—if not a ban—seems justified for good ethical and political reasons. Christof Heyns, during his time at the United Nations in the telling role of special rapporteur on extrajudicial, summary, or arbitrary executions, has raised concerns about the protection of human life as well as the preservation of "a minimum world order" given "the problematic use and contested justifications for drones and targeted killing." He argues that the onus is on those who want to deploy lethal autonomous robots to provide “considerable proof” that specific uses are permitted in specific circumstances.[17]

Killing at a Distance

Third, killing at a distance can be seen as morally problematic since psychologically speaking, it is easier to kill at a distance. Dave Grossman, a former West Point professor specializing in the psychology of killing in war, has contended that killing someone at close range, such as with a knife or one’s bare hands, is difficult because there is physical and empathic proximity, and the (fear of) trauma afterward.[18] When the distance increases, like in the case of artillery or bombing, killing becomes easier. The worry is that with drone fighting, remote killing becomes like a video game, making it all too easy to kill.[19] Military personnel hold that drone fighting is not like a video game, and similar procedures and decision-making are used as in the case of conventional, staffed aircraft. Also, drone killing involves not just the pilot or person pressing the button; there is an entire team of drone operators involved and of course a commander (present or remotely present) who takes decisions. That being said, the distance created and mediation by specific technology (a screen and joystick) remain, and they are likely to have at least *some* impact on how easy it is to kill. The worry is that the physical distance also impacts the moral distance—that phenomenologically, the “screenfighting” makes the face and body of people disappear, and thus dehumanizes people and removes barriers to killing them.[20] This reasoning is applicable not only to drones but all kinds of long-range weapons too, such as missiles and bombs, where there is a large distance between operators and targets. Think about the dropping of the atomic bombs in Japan at the end of World War II; those who dropped the bombs did not see the precise damage and concrete suffering on the ground. They did not see the death and suffering of the people below. Instead, from their godlike position high up in the air, they saw a mushroom-shaped cloud and felt the shock wave of the explosion. This distance seems to render empathy impossible. Some of this "moral distance" effect may also occur in the case of drone fighting at a distance.

Interestingly and tragically, however, operators have a much better view of what happens on the ground and to people in contemporary drone warfare. They see what people are doing and how they live their lives. Consider again the description in the beginning of this chapter, or these observations from another former drone soldier:

> Whenever I read comments by politicians defending the Unmanned Aerial Vehicle Predator and Reaper program—aka drones—I wish I could ask them a few questions. I'd start with: "How many women and children have you seen incinerated by a Hellfire missile?" And: "How many men have you

The worry is that the physical distance also impacts the moral distance—that phenomenologically, the "screenfighting" makes the face and body of people disappear, and thus dehumanizes people and removes barriers to killing them.

> seen crawl across a field, trying to make it to the nearest compound for help while bleeding out from severed legs?" Or even more pointedly: "How many soldiers have you seen die on the side of a road in Afghanistan because our ever-so-accurate UAVs were unable to detect an IED [improvised explosive device] that awaited their convoy?" . . . I knew the names of some of the young soldiers I saw bleed to death on the side of a road. I watched dozens of military-aged males die in Afghanistan, in empty fields, along riversides, and some right outside the compound where their family was waiting for them to return home from the mosque.[21]

Such descriptions suggest that the remote technology does not hinder but instead makes possible empathic experiences. Cameras and other technology linked to the drone create a new kind of "intimacy," in spite of the distance.[22] The "target" now appears as a spouse, parent, person making their way home, and so on. The "target" also gets a name. This "distant intimacy" and rehumanization through refacing and reembodiment of the "target" creates a psychological burden on the part of those who experience it.[23] Drone operators involved in lethal action are reported to experience stress; they not only have to make heavy decisions (e.g., is this a group of women and children, or are they combatants?) but also see what happens on the battlefield.[24] Being exposed to the graphic violence of destroyed homes and human remains, even these "drone warriors" who fight from a safe distance have found themselves traumatized.[25] Keeping in mind Müller's utilitarian argument, *these* consequences must be counted too.

This distant intimacy, or what we may call "remote proximity" phenomenon, and its psychological consequences is not only a bad thing; perhaps it could mitigate the problem that it is easier to kill from a distance, since in principle it could make killing harder rather than easier. It is nevertheless plausible that there is still a significant difference in experience between seeing death on the screen versus in front of your eyes when present on the battlefield. After all, the experience described by these ex-soldiers did not immediately stop them from contributing to drone killing. They continued to perform their dark duty. In any case, there remains a difference between war by aircraft and "boots on the ground." Soldiers on the ground come in much closer contact with the grim results of their lethal actions, and this functions as a natural psychological barrier against totally unrestrained killing, whereas the technology of drone fighting at least opens up that possibility; there is distance and (hence) no risk on the part of the attacker. As George Monbiot reminded us in one of his opinion pieces, military strategist Carl von Clausewitz thought that usually war does not escalate to absolute brutality because of "the risk faced by one's own forces."

The concern is that without risk, there is less restraint to kill. Monbiot concludes that "with these unmanned craft, governments can fight a coward's war, a god's war."[26] This also leads me to the next concern: Is this kind of "warfare" fair and virtuous?

Fairness and Virtue

Fourth, one could question if the asymmetrical surveillance and vulnerability relation created by the distance is fair. The drone operators see the "target" on the ground, but the person targeted may not know that they are being watched and are vulnerable to whatever the drone operator decides to do. The operator is not at risk, whereas the "target" carries all the risk. This "deus ex machina" situation, which gives the drone operator a godlike view and godlike power, is reminiscent of surveillance in a panopticon. The panopticon was a building design created by philosopher Jeremy Bentham in the eighteenth century, and is often referred to in contemporary discussions about power and surveillance. When used in prison design, the panopticon allows guards to see the prisoners, but the prisoners cannot see the guards; the prisoners do not even know if they are being watched. Drones create a similar situation. But in this case, the panopticon guards also shoot. This asymmetrical situation in terms of surveillance, control, and risk—which is not limited to drones, but also happens with other ways of remote fighting—could be seen as an unfair fight and even cowardly on the part of those who deploy the drones, going against the ethical principle of fairness *and* against military honor and virtue. Christian Enemark has argued that the risk-free use of drones challenges traditional notions of what it means to be a warrior, and raises questions concerning virtue and heroism. Is such warfare (still) more virtuous than organized murder?[27] The fact that many people who perpetrate these drone wars may not be concerned with fairness or virtue, and that there may be clear military and commercial interests at play here (e.g., funding the weapon industry), does not change the unethical character of this use of technology; arguably, it only makes things worse.

Explainability and Responsibility

Fifth, if drones use AI to make their own decisions, this raises the problem that some kinds of machine learning are unexplainable from a human point of view and thus drones' behavior is unpredictable. This is problematic when questions are asked before, during, and after the killing. How is responsible military action possible under these circumstances? The challenge of responsibility with regard to automation technologies is not only that—picking up Sparrow's assertion again—it is hard to hold anyone or anything responsible when we are concerned with highly automated systems such as drones, since machines lack any *human agency* and are out of our control (consider again the responsibility gap mentioned earlier). The trouble is also that what we may call the human moral *patients* may rightly ask for explanations from those who deploy the autonomous systems as to why and how decisions were made. Here the notion of responsibility is again understood in a relational way and as answerability; the question of responsibility is not only about agents along with their actions and knowledge as such but about being responsible and answerable *to* others too. Here these others include noncombatant civilians as victims or potential victims of drone attacks (and their political representatives), and perhaps even enemy combatants, all of whom may rightly demand answers when people are killed. They have the right to know how the decision was made. But in the case of AI-powered killer drones and other highly "intelligent" as well as automated military technological systems, neither the machine nor the commanding officer may be able to give satisfactory answers—the former because it cannot answer in a human sense (answer as a fellow human and give the kind of answers that humans need), and the latter because they do not know how the machine came to its decision. If this is the case, then it seems safer to limit the degree of automation, and keep humans in or on the loop in systems that they understand: have *them* take or supervise the decisions, and hence have them carry the responsibility, understood as the ability to *answer for* what they do. We need meaningful human control, but also effective human control and explainable systems that enable meaningful answering and answerability.

Should We Allow Machines to Kill Human Beings?

Finally, even if there was such a thing as a just war and even if the previous problems would not surface, one could argue that it is in itself morally problematic to have machines kill human beings. One could make a straightforward deontological claim that robots should never kill human beings, such as by drawing on the first of Asimov's laws of robotics that *robots may not injure a human being*. But one could also offer the following three reasons, which are related to discussions about what smart machines can or cannot do. First, machines may not only be less good than humans, *technically* speaking, at discriminating between targets and nontargets, combatants and noncombatants, and so on (which is, again, a ius in bello problem); as said, the decisions and actions of the machine may be *unpredictable*. Do we want unpredictable decisions regarding life and death? This was, among other things, precisely what Arkin-type views wanted to avoid by having "unreliable" human beings replaced by machines. It may be that humans cannot be trusted, but advanced machines that use AI may be unreliable in the sense of unpredictable. Of course this is an empirical matter; this may well change in the future. Second, even if they were technically good enough or even superior to humans, these machines would still lack a sufficient degree of moral *agency*. In particular, they would lack the capacity for ethical decision-making about life and death. For example, one could contend that the lack of emotions is not a pro, as Arkin and Müller assert, but rather a contra; if emotions are part of good human ethical judgment and can contribute to more ethical action, then we need emotions, and therefore need to keep the human in the loop. Empathy, for instance, may protect against unrestrained video game–style killing—if this is an adequate description at all. Third, even if the machines were given some machine morality and even if there was no principled objection to this kind of morality, they nevertheless lack humanlike moral patiency and therefore should not be allowed to kill humans. Humans know what it means to fear death and experience their vulnerability. If an entity is not alive, and lacks the experience of human existence and vulnerability, then arguably it should not be allowed to take the life of a human being, or make decisions about the life and death of human beings.

Both the lack of full moral agency and the lack of full, humanlike moral patiency thus create fundamental asymmetries between robots and machines, and constitute good arguments against fully automated killing.[28] As Heyns put it in his UN report, "Machines lack morality and mortality, and should as a result not have life and death powers over humans."[29] Killing human beings, if ever justified, should be left to humans.

If an entity is not alive, and lacks the experience of human existence and vulnerability, then arguably it should not be allowed to take the life of a human being.

During the past years, and partly but significantly as a result of activism and lobbying by civil society groups and experts, there is increasing support for this position at the international level. UN secretary-general António Guterres told a meeting of experts in lethal autonomous weapon systems that "machines with the power and discretion to select targets and take lives without human involvement are politically unacceptable, morally repugnant, and should be prohibited by international law."[30] The European Parliament has called for negotiating an international ban. Some countries already took their own initiatives. For example, the Belgian Parliament's defense committee adopted a resolution that asks the government to support an international ban.[31] And some companies and an increasing number of technology workers refuse to develop these systems. Yet major global military and political players continue with their plans, and develop and/or use UAVs with lethal capacity.

Discussing the ethics of killer drones thus makes us reflect not only on autonomous technology but also on the justification for war as well as what humans do and should (not) do in war, killing and distance, military virtue, and global politics. Along the way, different kinds of moral arguments and moral theories were used. For example, Arkin and Müller use consequentialist reasoning to defend the use of such machines, whereas many of the arguments contra were based on deontological views about what kind of actions are morally permissible (and there was even one virtue ethics argument about cowardliness). In the end, the discussion has led us to consider a philosophical view of human being and human existence that takes seriously the moral significance of human mortality and vulnerability for thinking about killing. It is a view that together with all kinds of ethical and legal principles, such as bioethics principles and human rights, could form the basis for an ethics of killer drones—or perhaps one should say, an ethics that is not about machines and *their* morality but instead aimed at the protection of *humans* and human *lives*.

8

ROBOTIC MIRRORS BEYOND THE HUMAN: ROBOT ETHICS AS AN ENVIRONMENTAL ETHICS

As organizers of the Robophilosophy 2018 conference in Vienna, we asked participants, Why do we build robots and why should we build robots at all?[1] Many reasons were offered: we build robots as the means to reach economic ends, do dirty or dangerous work for us, advance our scientific understanding, satisfy human needs such as sociality, project and externalize our hopes and fears, satisfy our narcissist desires, create a better world or transcend the human condition. Robots are also part of historical, social, cultural, and linguistic contexts and processes. There is always a "who" that builds robots—we have to ask questions about society, politics, and power too—and we construct robots with words, tell stories about robots, and even talk to robots. In addition, robots express norms and threaten or help to realize ethical values.[2] But one of the most interesting, well-known reasons given was that robots help

Figure 7 Still from the film *Ex Machina*. Courtesy of Universal Studios Licensing LLC.

us to understand what makes us human. This is what this chapter is about. Questions about robots lead us to questions about humans and humanity. What does it mean to use robots as mirrors of humankind, what could possibly be problematic with this project, and what are the implications for robot ethics?

Robotic Mirrors: All about Humans

Robot ethics and, more generally, the philosophy of robotics are not only about robots but very much about humans too. As the chapters in this book show, robot ethics is of course about machines, but it is also concerned with a range of themes that are all about the human: what meaningful work is, how our societies are transforming and how they should be organized, what good social and personal relationships are, if it is acceptable to deceive people, what it means to respect human dignity, what good health care is, what ethics and morality are, what we mean by moral agency and responsibility, whether we owe anything to nonhumans, how the moral standing of humans themselves is grounded, what a just and morally acceptable war is (if it is acceptable at all), and how the experience of human vulnerability and mortality can function as a basis for ethics. Robot ethics is thus about us: the present and future of our morality, societies, and human existence. In this sense, robots are not only object of our ethics and other philosophical work, as something that we think *about* but that does not further touch us; rather, robots function as mirrors that show and reflect *us*—that is, the human being in all its facets, and with all of its problems and challenges, including ethical ones. Robot ethics is always connected to philosophical anthropology. We use robotic mirrors as a philosophical-anthropological tool.

The idea of using machines as mirrors for philosophical reflection on the human is not new. René Descartes already did this when he argued that animals are complex machines—much better made than our machines because they are made by God, yet still "just a machine"—but that humans have distinctive features. We have animal bodies, our bodies are machines, and yet in addition we have reason. A machine may utter words, but we can give

Robots function as mirrors that show and reflect *us*—that is, the human being in all its facets, and with all of its problems and challenges, including ethical ones.

meaningful answers to whatever is said and use reason as a universal instrument in all kinds of situations. Hence, whereas we may not be able to tell the difference between an animal and a machine shaped like an animal and with the organs of an animal, we would always recognize real humans.[3] More generally, we have always used other entities to think about ourselves, and in fact to define ourselves as humans and distinguish ourselves from these other entities. Consider animals, like in Descartes's argument, but also god(s), angels, demons, golems, homunculi, machines, monsters, zombies, aliens, and indeed robots and AI.

As the example of Descartes's anthropology shows, often these exercises have constituted what I have called "negative anthropologies"; we tend to define the human in terms of what it is not.[4] We are not x: we are not animals, we are not gods, and so on. Similarly, *technoanthropologies* that use robots are usually negatively formulated: we are not robots, we are not machines, and we are not entities with AI. But this negative approach also means that apparently in modernity, we *need* robots to define the human. They play a key role in our negative anthropologies. Whatever else robots may be, they are *technoanthropological* tools. We only achieve the distinction through constructing contrast. The robotic mirror gives us a negative image, which enables us to delineate the human. We are nonrobots, nonmachines. Or we are more than robots, more than machines (as in Descartes).

An important influence on this particular technoanthropology is humanism, broadly understood: a philosophical view that emphasizes human beings and their value. Humanist anthropologies as they developed in the West care about distinguishing humans from nonhumans, and if possible, maintaining and defending that distinction. We develop Turing and other tests along with thought experiments to make sure we can do this. Other cultures seem less obsessed with establishing and maintaining such distinctions. For example, influenced by traditional Shinto and Buddhist worldviews, Japanese culture seems to care a lot less about the question of what makes the human special.

In robot ethics, this negative technoanthropology and humanism frequently takes the form, normatively speaking, of defending the human against the machine. Humans and robots are seen in an antagonistic framework. The relation between humans and robots is defined by a competition narrative. We say that they should not take our jobs. It is either the human or the machine. Robots are our tools; they should not become our masters. Human values should prevail. Based on an anthropology that tries to sharply delineate the human from the nonhuman and puts the human first (or highest), ethics—and therefore robot ethics—is human-centric and defensive when it comes to what is perceived as a threat. Robots are seen as challenging us to a duel for our distinctiveness and standing. Perhaps they already attack. But this human-centric and defensive approach is not the only option, anthropologically and ethically speaking.

Beyond the Merely Human

What happens to robot ethics if we move beyond a human-centric and negative anthropology and ethics? Should the mirrors become open windows, and if so, how and what are the consequences for robot ethics? I already mentioned this issue in chapter 3, but it deserves further discussion. There are at least the following options for a robot ethics, and more generally, an anthropology and ethics, to go beyond the merely human.

First, there is the option to not resist "the machines" and rather embrace that humans are *a kind of* robot. We are the robots, as a famous Kraftwerk song proclaims. In this option, we accept that we are machines, albeit biological ones. This modern-scientific, rather Cartesian anthropology implies that we can also improve ourselves—as we already do, such as by means of medicines and prosthetics; we are already cyborgs. Nature made us limited machines; we can do better and upgrade ourselves—that is, if we are necessary in this universe at all in the long run. Maybe it is better to be replaced by highly intelligent machines: robots with AI that outsmart us. We can make machines that are (or become) better than us. This is, in a nutshell, at least one, influential version of transhumanism. Resistance against the machines is futile since we and our humanism are already outdated. Through genetic modification and techniques such as gene editing, but with the help of all kinds of material technologies too, like brain-computer interfaces and artificial body parts, we can and should "enhance" ourselves. In this way, we become a kind of robot, cyborgs—combinations of biological and nonbiological materials—and perhaps we will eventually be greeted or replaced by what roboticist Hans Moravec has called our "mind children." If our machines become as complex as human beings, they will be our descendants. As Moravec asserts, "Unleashed from the plodding pace of biological evolution, the children of our mind will be free to grow to confront immense and fundamental challenges in the larger universe."[5] Instead of fearing that robots will take over, Moravec suggests, we should look forward to a wonderful postbiological future in which we can participate.

But how will we get there, and what will be our place? Contemporary transhumanists have different views about this, and some are more worried about the potential risks than others. Ray Kurzweil has argued that we will become cyborgs and ultimately upload ourselves.[6] We can then inhabit a digital sphere and/or get embedded in a machine. His vision is generally optimistic. Nick Bostrom writes about machines that surpass us in intelligence, but is also concerned with what this might mean for us humans; he raises the problem of how to control what a superintelligence would do (the control problem).[7] *If* this development were to occur, our fate as a species would depend on the actions of such a superintelligence. Max Tegmark thinks that there will be an intelligence explosion and intelligent life will then spread in the cosmos. If this happens, we will be humbled by ever-smarter machines. He acknowledges risks, but at the conclusion of his book points out that we can improve the prospects of a happy ending to the current revolution "if we can create a more harmonious human society characterized by cooperation towards shared goals" and take up our responsibility as "guardians of the future of life," shaping that future now and together.[8]

In light of these transhumanist visions, doing robot ethics can mean building machines that have morality or are smart enough to figure out morality themselves, perhaps inventing a new kind of morality that is superior to that of humans, who are unnecessarily hindered by limitations such as emotions. One may then hope that machines can realize the best, most rational ethics. Instead of always focusing on human ethics, as humanism does, we should create conditions under which machine morality can advance. Here robot ethics means machine ethics. With rational thinking and thought experiments such as morality games, philosophers can help to establish such an ethics, together with scientists and robotics researchers.

Moreover, philosophers can help by thinking about the control problem and other implications for the future of humanity. Robot ethics then means guiding the development of machines that will benefit humanity. In *Superintelligence*, for example, Bostrom discusses some methods and techniques we could use for avoiding existential catastrophe. And Tegmark's suggestion could be interpreted as a call for ensuring that the further development of robotics and (other) intelligent technologies is embedded in efforts that aim at transforming our society in a harmonious, cooperative direction.

Second, one could take a posthumanist route, which is not so much focused on what new, exciting, super smart machines we can build and what existential risks they pose for humanity but rather embraces robots as part of its celebration of hybridity, boundary crossing, and promiscuous relating to other entities. In this vision, humans, animals, and robots can and should live together. Instead of replacing the human by robots, as humanism fears and transhumanism looks forward to, the idea is that humans and nonhumans such as robots relate and merge. Robots can be our friendly artificial others, or we can explore hybridity with machines. We can get closer to them or perhaps we have always been close to them. What the human "is" (there is no human "essence"), is and should not be clearly delineated. We should move beyond dualist and binary categories. Cyborgs figure again, but this time not so much as material combinations of biological and nonbiological machines that should upgrade us as part of a hypermodern project but rather as symbols of the postmodern crossing of boundaries and the liberation from binaries. Machines are then not unwelcome immigrants that want to take our place (as humanism fears) or welcome children that we have to release into the cosmos when they become far more intelligent than us (as transhumanism desires); they are a potential partner for postmodern play as well as an object on which we can project our psychological desires, ethical aspirations, and political hopes. It is time for robot love.

Donna Haraway is an important influence on this version of posthumanism. In her famous "A Cyborg Manifesto," Haraway offered the cyborg metaphor as a tool to question boundaries between humans, animals, and machines, leading us to a world in which we are no longer afraid of "joint kinship" with animals and machines, and in which we accept monsters that break down barriers. We should not fear couplings between organisms and machines, and if robots seem alive, this is only a problem if we hang on to a Western and Cartesian dualism.[9] Robots, in the form of robots that seem alive or as cyborgs, could in this vision play a role in liberating us from unnecessary boundaries that still figure in our modern thinking and doing. They can join and help to establish a postmodern utopia where binaries are gone and there is symbiosis—a living together of humans, animals, and machines. Or perhaps together we already inhabit, à la Bruno Latour, a nonmodern world, where the social consists of humans and nonhumans, and there is hybridity everywhere.[10] In such a vision, there is no need for posthumanist fantasies since we have never been modern in the first place. Modern philosophy asks us to distinguish between subjects and objects, humans and machines, people and things. But we have always lived in hybridity. Robots are already part of the social.

Rosi Braidotti is another contemporary voice in posthumanism. Also in a postmodern way, she interprets the posthuman in terms of multiple identities and argues for postanthropocentric thinking. Like Haraway, she believes that the categorical distinction between humans and other species should be erased. Braidotti discusses the blurring of human/machine boundaries too. In *The Posthuman*, she argues against a categorical distinction between nature and culture, and proposes "a non-dualistic understanding of nature-culture interaction" given that science and technology have blurred the boundaries. The complexity of our increasingly smart and autonomous technologies lies for her at the core of the postanthropocentric turn. Influenced by philosopher Félix Guattari (who builds on biologist Francisco Varela), Braidotti claims that both biological organisms and machines are autopoietic (self-organizing). As Haraway already observed, machines become more alive, machines become immaterial, and bodies become connected to machines. Again we meet the cyborg, but Braidotti includes "the anonymous masses of the underpaid, digital proletariat who fuel the technology-driven global economy" as well in her posthuman imaginary, and argues for thinking in terms of a "radical relationality" that merges the subject with its "technologically mediated planetary environment," establishing a new kind of "symbiotic relationship" and ethics of mutual interdependence. She rejects, however, "the extreme forms of science-driven post-humanism which dismiss the need for a subject altogether"; this can be interpreted as an attempt to maintain distance from (some forms of) transhumanism.[11]

From a posthumanist point of view, robot ethics could then mean that we find a good way of relating to, and living together with, machines, and develop, use, and think about robots in a way that contributes to the breaking down of dualist and binary thinking as well as practices, and embraces hybridity and monsters. Such an ethics does not defend the human against the machine but rather thinks about how to shape human-robot couplings and construct a hybrid human/nonhuman society. Art and fiction can help us to explore this. Moreover, inspired by Braidotti's critical posthumanism, we could go beyond robot ethics understood as the ethics of human-robot interaction and consider the political economy of the cyborg. Could robots and robotics be part of a transformed socioeconomic system that is no longer exploitative? What would it mean to see robots as part of a technologically mediated planetary environment? How can robotics help to reach a new, post-human symbiosis of human and nonhuman, nature and culture, technology and biology?

Third and finally, going beyond the human can point in the direction of the natural environment, and suggest a more ecological, less human-centered "anthropology" and ethics. While this approach could in principle liaise with the previous ones (e.g., with Latour or Braidotti), in this option the discourse is less technophile than either transhumanism or posthumanism. The focus, anthropologically and ethically speaking, is on the natural environment, planet, or earth. One could then argue that instead of being mesmerized by transhumanist science fiction and posthumanist fantasies about cyborgs, we should focus on real and urgent problems with the natural environment and planet like climate change. If we want to enhance anything at all, we should not forget that we totally depend on the earth and nature for any technological development; we better try to improve *that* relationship and *those* conditions. And if we want to become more relational, we should start with caring about the natural environment, the ecologies of which we are a part.

In this vision, robot ethics should not only care about humans (as humanists do) or robots (as in transhumanist or posthumanist visions). The development and use of robots, if needed at all, should contribute to the realization

Instead of being mesmerized by transhumanist science fiction and posthumanist fantasies about cyborgs, we should focus on real and urgent problems with the natural environment and planet like climate change.

of a natural environment and planet that is ecologically sustainable, and not undermined or destroyed by humans and their technologies. And this is important not only because other living beings and natural environments have inherent worth and value, next to human beings; even from an anthropocentric point of view, we should work toward a sustainable world, including a livable climate on earth. It might well be the only way to *survive*, and it is certainly the only way to (begin or continue to) live well and flourish. Furthermore, humanism is not the only culprit. *Neither* transhumanist and modern scientific efforts to manipulate nature and escape the planet *nor* posthumanist visions of nature-culture hybridity are sufficient to ensure that there is a future for humans and other living beings. If robotics and robot ethics continue their obsession with human goals and values, or remain solely focused on dreams of a perfect technological future (with or beyond the human), they will contribute to the corruption and destruction of the ecosystem(s) of the only planet we have—earth—and thereby in the end to the extinction of many species including perhaps the human one. Instead, we need a robot ethics that is centered on the environment. To conclude, in robot ethics we need to pay not only more attention to how "environmental robots" can help us to respond to "unprecedented environmental changes due to the human influence on Earth systems" as well as the ethical and political issues they raise;[12] robot ethics itself should be transformed and become an *environmental ethics*.

What does this mean? What does it mean to conceive of robot ethics as an environmental ethics? I propose to distinguish between a narrow (or weak) interpretation and a broad (or strong) one. The weak interpretation is that we need robots that are environmentally friendly in the sense that, put negatively, they minimize or (preferably) avoid harm to the natural environment and its living beings, and positively, help us to respond to environmental challenges such as climate change. (These are the environmental robots that Aimee van Wynsberghe and Justin Donhauser call for.) To reach these goals in and with robotics would already be an enormous step forward, seen from an environmentalist and ecologist point of view. I expect that this goal will receive, or is already receiving, much support from people in the robotics community and beyond. During the past decades, modest forms of environmental and ecological thinking have become more or less mainstream. Many researchers are already interested in, or working on, environmentally friendly and sustainable robotics. The strong interpretation, by contrast, puts the very concept of "robot" and its technological futures into question. What kind of artificial creatures do we need, if we need them at all, given the environmental predicament we are in today? Or do we need them because the human ought to be superseded? What is the future of robotics in light of a radically relational and environmental vision?

Answers to these questions may depend on which position is taken in relation to the other mentioned approaches. A transhumanist, if inspired by environmentalism, and not arguing for abandoning the earth and colonizing other planets, may call for humans to be enhanced and eventually replaced by, or merged with, smarter creatures that will most likely do a better job at managing the planet. In this scenario, the robots will take over in order to reach an environmentalist goal: superintelligent machines will save the planet, although it may not be worth saving humanity as it is. Critical posthumanists and environmentalists will go in other directions, perhaps welcoming artificial creatures next to natural ones and inviting them to cohabit this earth, or questioning key modern and transhumanist assumptions such as anthropocentrism and its consequences for the natural environment and the planet.

Here are *my* questions, inspired by the latter approaches. Could it be that embracing robot ethics as an environmental ethics has to lead to abandoning the very project of modern robotics? If robots are tools for gaining total control over the earth, and if that is problematic, then is it enough to build environmentally friendly robots, or do we need more radical changes, maybe not involving "robots" at all? If robots are slaves to serve us, but if social relationships and patterns of slavery, domination, and the hegemonic relations they are embedded in are morally reprehensible, then do we need "robots" at all? If robots mirror the human, but that "human" assumes the ugly face of hegemonic subjectivity along with the totalitarian control of people and nature, do we need that kind of mirror? Posthumanist and environmentalist literature offers suggestions for alternative ways of ethical and political thinking—ways of thinking that go beyond old-style humanism and that reject strong anthropocentric and dualist modern views of nature. They also offer approaches that go in new *political* directions, questioning modern power structures. More work is needed on the politics of robotics.

While it is far from clear what these approaches and this strong interpretation of robot ethics as environmental ethics imply for the future of technology and robotics, it seems to me that a critical and philosophical robot ethics should not avoid exploring these directions, and asking and discussing these questions. For example, if value sensitive design is the right kind of direction for robotics, then why only take into account human values? Why should design center on the human at all? Is it sufficient to *also* take into account values of nonhumans and principles such as sustainability, or can we conceive of a more radically environmental and politically critical kind of design? And—to stay within our topic—would the outcome of such design still count as a *robot*? What would an environmental "postrobot" be or look like?

If robots mirror
the human, but that
"human" assumes the
ugly face of hegemonic
subjectivity along
with the totalitarian
control of people and
nature, do we need
that kind of mirror?

The sketches of philosophical positions in this chapter are somewhat caricatural, but this is done to bring out more clearly the differences between the various directions (humanist, transhumanist, posthumanist, and environmental) and invite readers to think about the possibilities that these approaches may offer for the further development of robot ethics. Once again, we see how robot ethics is not just about robots, machines, or artifacts; it concerns the "big" questions about humans and—to the extent that we need to, or can, go beyond humans in our thinking and ethics—in any case the big questions *for* humans that can only be answered *by* humans. Robots may help us with this exercise, as tools and a mirror for our thinking about these questions. As we have seen, the mirror shows several, distinct images. But in the end, we humans have to interpret, imagine, negotiate, and decide, as persons, societies, and perhaps humanity on this planet, which direction we go and should go. In other words, we humans have to think about robots, but also about the arguably more important matters raised by robots, robotics, and human-robot interaction. This thinking cannot and should not be left to machines. Ethically and politically speaking, humanism might be problematic, at least in the senses criticized by transhumanism and posthumanism. Epistemologically, however, it makes sense if it means that we recognize that we always have to go via the human and human subjectivity when we think—about robots, robot ethics, and other matters. We do not just "use" the mirror; we are part of the *mirroring*.

Finally, it should be asked *which* humans or *who* (individuals, groups, or categories of people) should engage in robot ethics and these mirroring exercises, apart from academics doing robot ethics and robot philosophy on a professional basis. One answer is roboticists. As I suggested in the first chapter, engineers, developers, and anyone involved in the design and making of robots have a significant responsibility with regard to their creations. To enable them to exercise this responsibility and engage in ethical reflection, it is crucial to build ethics into the curriculum of technical educational programs and trainings—not as an add-on, but integrated into all relevant technical courses. Efforts should also be undertaken to create more diversity (e.g., in terms of gender) with regard to the student population and composition of teams doing robotics research. Moreover, in line with ideas around responsible research and innovation, and the analysis of responsibility in terms of many hands, many others should participate in doing robot ethics, such as users of robots, of course, but other stakeholders too, such as investors and shareholders, executives and other managers in robotics companies, consultants, professionals and practitioners in a particular field (e.g., care or education), governments, and nongovernmental organizations like consumer rights organizations and workers unions. If they make or influence decisions about robotics, or bear the consequences of robotics, they should also reflect on its ethics and politics. And if it is true that in the end, robot ethics—like any technology ethics—concerns us all as members of a community, citizens of a particular society, and members of humanity, if we are all stakeholders to some extent, then we should all think about, have a say in, and take some responsibility for it. Let the robotic mirror show us how democratic we really are, can be, and want to be when it comes to the future of technology and society.

ACKNOWLEDGMENTS

I wish to thank my editors at the MIT Press, Philip Laughlin, Alex Hoopes, and Virginia Crossman, for supporting this book project. I warmly thank my colleague Zachary Storms for helping with formatting the manuscript and sorting out the copyrights for the images. I also thank my friends and family for being there for me—online and off-line, and during difficult times. Finally, I am grateful to my awesome friends and colleagues in the robot ethics community; it was a pleasure to engage in discussions and spend time with them over the past fifteen years, and the field would not be what it is today without them. Let's keep the good work going!

GLOSSARY

Androids
Robots designed to appear like humans.

Anthropomorphizing
Attributing human characteristics to nonhuman entities such as gods, animals, or objects. In robotics and human-robot interaction, the term refers to the phenomenon that humans project humanlike properties such as feelings onto a robot.

Artificial general intelligence
The hypothetical intelligence of a machine that can understand or learn any cognitive task that human beings can.

Asimov's laws of robotics
A set of ethical rules for robots offered by science-fiction writer Isaac Asimov. The three laws are: a robot may not injure a human being or, through inaction, allow a human being to come to harm; a robot must obey the orders given to it by human beings except where such orders would conflict with the first law; and a robot must protect its own existence as long as such protection does not conflict with the first or second laws. Later, law zero was added: a robot may not injure humanity or, by inaction, allow humanity to come to harm.

Capability approach
Originally an economic theory, it now refers to a normative conceptual framework that understands people's freedom to achieve well-being in terms of their capabilities—that is, what individuals are capable of doing.

Capitalism
An economic and political system based on the private ownership of the means of production and free market competition, which according to Karl Marx and other critics, leads to capital accumulation and power in the hands of a few along with the exploitation of workers.

Cyberphysical systems
Technological systems that integrate computation, networking, and physical processes. The system is controlled by algorithms and connected to the internet.

Environmental robots
Robots that help us respond to environmental challenges, including the problem of climate change and its consequences for humans and the earth's ecosystems.

Ethics
A subfield of philosophy concerned with normative questions about the right (behavior) and the good (life).

Fetish
A term borrowed from Freudian discourse, where it refers to a specific form of sexual pathology, it is used here in the more general sense of a (desired) human-made object that is a surrogate for something else.

Fourth (industrial) revolution
The idea that after the three previous industrial revolutions, each of which were enabled by new technology (e.g., the steam engine), industry is now transformed by the use of autonomous and intelligent robots as well as (other) cyberphysical systems and the Internet of Things.

The good life (*eudaimonia*)
A philosophical term; philosophers have different views about what a good life means. In the Aristotelian tradition, it is connected with achieving personal excellence or virtue.

Humanism
A philosophical view that centers on human beings and their value.

Industry 4.0
Originally used in the context of German industry policy, the term now refers to the idea that we are moving toward smarter manufacturing and smarter factories, using an Internet of Things and (big) data analytics next to other new technologies such as 3D printing, cloud technology, augmented reality, and so on.

In the loop versus on the loop versus out of the loop
Distinctions concerning how involved humans are in the control of autonomous systems. Do humans have direct control and are their actions needed to make the system act (in the loop), do they only supervise (on the loop), or is there no interaction with the system since it works autonomously (out of the loop)?

***Ius ad bellum* versus *ius in bello* (also, *jus ad bellum* versus *jus in bello*)**
The distinction between the question regarding the justification of (starting a) war and the question of just conduct in and during a war.

Killer drones
UAVs (see below) with lethal capacity—that is, the capacity to kill.

Leisure society
The utopian idea of a society in which machines have taken over work previously done by humans, liberating those humans to live a life of leisure.

Moral patients
Entities toward which moral agents can have moral responsibilities. They are on the receiving end of the moral relation and something is due to them.

Posthumanism
A range of philosophical views that expand the circle of philosophical and ethical concern to nonhumans, and positively evaluate boundary crossings and mergings, such as by using the metaphor of the cyborg.

Responsibility gap
In the ethics of robotics and automation, this gap refers to the problem that robots and autonomous systems have increased autonomy and agency, but no responsibility. Responsibility can then be ascribed to humans, but who is responsible, and can humans still intervene on time?

Responsible research and innovation
Research and innovation processes in which societal stakeholders and innovators are mutually responsive toward each other, and in which ethics is integrated from the beginning.

Robot

A robot is an autonomous machine capable of sensing its environment, carrying out computations to make decisions, and performing actions in the real world (based on the definition of the Institute of Electrical and Electronics Engineers). This book focuses on robots that in addition to this definition, have hardware as well as a high degree of autonomy, intelligence, and interactivity.

Robot ethics

A field of the philosophy of technology and applied ethics concerned with normative questions regarding robots. It can refer to how humans who use and develop robots should behave (e.g., toward other humans or robots), or how robots should behave.

Robot philosophy

Philosophy concerned with questions about robots.

Socially assistive robots

Robots that can help people in contexts such as education, training, therapy, and rehabilitation through social interaction.

Social robots

Robots designed for imitating human social behavior in order to facilitate human-robot interaction.

Superintelligence

A hypothetical agent that has intelligence far exceeding that of the smartest humans.

Suspension of disbelief

Suspending one's critical faculties and believing something for the sake of enjoyment.

Transhumanism

The belief that humanity can and should evolve beyond its current limitations by means of science and technology.

Trolley dilemma
A thought experiment in ethics that presents the following problem or some variant: a trolley is rolling toward five people tied to a track, and you control a lever, which if pulled, redirects the trolley onto another track where there is a single person lying on the track. Do you do nothing or pull the lever?

Turing test
A test proposed by Alan Turing. The aim is to test if a machine exhibits intelligent behavior indistinguishable from that of humans. In the test, a human evaluator judges natural language conversations between a human and a machine designed to provide humanlike responses. If the evaluator cannot reliably tell the machine from the human, then the machine has passed the test.

Uncanny valley
Hypothesis put forward by Masahiro Mori that familiarity increases with human likeness, but at some point there is an "uncanny valley," in which a subtle deviation from human appearance creates a feeling of strangeness or uncanniness.

Universal basic income
A periodic payment given by the state to all citizens on an individual basis without the requirement to work or report what one does.

Unmanned aerial vehicles (UAVs)
Aircraft piloted by remote control or onboard computers. One could also call it a "flying robot."

Wizard of Oz experiments
Research experiment in which subjects interact with a computer or robot that they believe to be autonomous and intelligent, but that is actually (remotely) operated by a human being.

NOTES

Chapter 1

1. Coeckelbergh, *New Romantic Cyborgs*.
2. https://robots.ieee.org/learn/.
3. Lin, Abney, and Bekey, "Robot Ethics," 943.
4. Heidegger, *Question concerning Technology*.
5. Coeckelbergh, "You, Robot."
6. Asaro, "What Should We Want from a Robot Ethic?," 9.
7. Coeckelbergh, "Robotic Appearances."
8. Abney, "Robotics, Ethical Theory, and Metaethics."
9. Botting, *Mary Shelley*.
10. Boddington, *Towards a Code of Ethics*; Coeckelbergh, *AI Ethics*; Gunkel, *An Introduction*; Liao, *Ethics*; Bartneck et al., *An Introduction*.
11. See, for example, Hildebrandt, *Smart Technologies*; Turner, *Robot Rules*; Fosch-Villaronga, *Robots*.
12. Lin, Bekey, and Abney, *Robot Ethics*; Lin, Abney, and Jenkins, *Robot Ethics 2.0*; Coeckelbergh et al., *Envisioning Robots*.

Chapter 2

1. https://libcom.org/blog/xulizhi-foxconn-suicide-poetry.
2. Wakefield, "Foxconn."
3. Marx, *Capital*, 548.
4. Wiener, *Human Use*, 162.
5. Servoz, *Future of Work*, 40.
6. Fletcher and Webb, "Industrial Robot Ethics."
7. UNI Global Union, *Top 10 Principles*, 4–5.
8. Zuboff, *Age of Surveillance Capitalism*.
9. Brynjolfsson and McAfee, *Second Machine Age*.
10. Schwab, *Fourth Industrial Revolution*.
11. Frey and Osborne, "Future of Employment."
12. McKinsey Global Institute, *Jobs Lost*.
13. McKinsey Global Institute, *Jobs Lost*.
14. Ford, *Rise of the Robots*.
15. Frey and Osborne, "Future of Employment," 20–22.
16. Servoz, *Future of Work*, 5.

17. McKinsey Global Institute, *Jobs Lost*, 39.
18. McKinsey Global Institute, *Jobs Lost*, 36.
19. PricewaterhouseCoopers, *Will Robots Really Steal Our Jobs?*
20. Servoz, *Future of Work*, 6.
21. World Economic Forum, *Reskilling Revolution*.
22. PricewaterhouseCoopers, *Will Robots Really Steal Our Jobs?*
23. PricewaterhouseCoopers, *Will Robots Really Steal Our Jobs?*
24. McKinsey Global Institute, *Jobs Lost*.
25. Servoz, *Future of Work*, 3.
26. Servoz, *Future of Work*, 45–46.
27. World Economic Forum, *Future of Jobs*, ix.
28. Danaher, *Automation and Utopia*, 135.
29. Veal, "Leisure."
30. McKinsey Global Institute, *Jobs Lost*, 8.
31. PricewaterhouseCoopers, *Fourth Industrial Revolution*.

Chapter 3

1. https://www.jibo.com/.
2. https://futurism.com/jibo-dead-announcement-video.
3. See the famous Google Duplex demo in 2018: https://ai.googleblog.com/2018/05/duplex-ai-system-for-natural-conversation.html.
4. For an overview of the discussion, see van Den Hoven et al., "Privacy."
5. Zuboff, *Age of Surveillance Capitalism*.
6. Véliz, *Privacy Is Power*.
7. http://hellobarbiefaq.mattel.com/.
8. Reese, "Wi-Fi-Enabled 'Hello Barbie.'"
9. Calo, "Robots and Privacy," 188.
10. Matsuzaki and Lindemann, "Autonomy-Safety Paradox."
11. Sparrow and Sparrow, "In the Hands of Machines?," 155.
12. Sparrow, "Robots in Aged Care," 448, 446.
13. Coeckelbergh, "Care Robots and the Future of ICT-Mediated Elderly Care."
14. Sharkey and Sharkey, "Crying Shame," 166, 168.
15. Turkle, *Reclaiming Conversation*, 344, 339, 358.
16. I write "loosely Freudian" rather than "Sigmund Freud's," since I keep distance from his 1927 paper "Fetishism" in which he outlines a theory rooted in his specific views about sexual development and related pathologies. In the paper, Freud sees fetishism as a male perversion linked to the child's fear of castration: confronted with the mother's lack of a penis, the fetishist looks for an object as a substitute for that lacking penis. In the 1950s,

Jacques-Marie-Émile Lacan further developed this theory. The meaning of fetish that I use here, by contrast, is much broader and more common; it refers to an object that is a surrogate for something else. The term "fetish" is thus used without any intention to evoke the specific theories of sexual pathology developed by Freud and Lacan. Furthermore, my use of the term "magic" here refers to use of the term "fetish" in the context of anthropology, but again without referring to specific theories. Elaborating this topic would distract from the introductory aim of this book.

17. Sullins, "Robots, Love, and Sex," 208.
18. Carpenter, "Deus Sex Machina," 283.
19. Smith, *Erotic Doll*.
20. Richardson, "Asymmetrical 'Relationship,'" 291, 292. For the campaign, see https://campaignagainstsexrobots.org/.
21. Isaac and Bridewell, "White Lies," 157.
22. For more discussion on this issue, see Coeckelbergh, "How to Describe."
23. Coeckelbergh, "Technology Games/Gender Games."
24. Carpenter et al., "Gender Representation," 262.
25. Carpenter et al., "Gender Representation," 264.
26. Benjamin, *Race after Technology*.
27. Coeckelbergh, "Technology Games/Gender Games."
28. Carpenter, "Existential Robot," 42.

Chapter 4

1. Foster, "Aging Japan."
2. http://www.parorobots.com/.
3. Archer, "'Friendly' Hospital Robot."
4. https://med.nyu.edu/robotic-surgery/physicians/what-robotic-surgery.
5. Mavroforou et al., "Legal and Ethical Issues in Robotic Surgery."
6. Nordrum, "Cyderdyne's HAL Exoskeleton."
7. van Wynsberghe and Li, "Paradigm Shift."
8. https://www.dream2020.eu/goals-methodology/.
9. https://www.softbankrobotics.com/emea/en/nao.
10. See, for example, Coeckelbergh et al., "Survey of Expectations."
11. Sparrow and Sparrow, "In the Hands of Machines?"; Sharkey and Sharkey, "Granny."
12. Matthias 2004, "Responsibility Gap."
13. Sparrow and Sparrow, "In the Hands of Machines?"; Sharkey and Sharkey, "Granny."
14. Kitwood, *Dementia*, 47. See also Sharkey and Sharkey, "Granny."

15. Sparrow and Sparrow, "In the Hands of Machines?"
16. Coeckelbergh, "Health Care."
17. Nozick, *Anarchy*, 43.
18. Sharkey and Sharkey, "Granny."
19. Beauchamp and Childress, *Principles*.
20. Sparrow and Sparrow, "In the Hands of Machines?"
21. Sharkey and Sharkey, "Granny."
22. Nussbaum and Sen, *Quality of Life*; Nussbaum, *Women*; Coeckelbergh, "Health Care," 184–185; Coeckelbergh, "How I Learned."
23. Nussbaum, *Frontiers of Justice*, 76.
24. Nussbaum, *Frontiers of Justice*, 76–78.
25. Coeckelbergh, "How I Learned," 79–80.
26. Borenstein and Pearson, "Robot Caregivers."
27. Coeckelbergh, "Care Robots, Virtual Virtue."
28. Sparrow, "Robots in Aged Care," 448.
29. Coeckelbergh, "E-Care as Craftmanship."
30. Sennett, *Craftsman*.
31. Dreyfus and Dreyfus, *Five-Stage Model*; Dreyfus and Dreyfus, "Towards a Phenomenology."
32. Coeckelbergh, "E-Care as Craftmanship."
33. Sorell and Draper, "Robot Carers," 188.
34. Coeckelbergh, "Care Robots, Virtual Virtue."
35. Coeckelbergh, "Artificial Agents," 267–270.
36. Coeckelbergh et al., "Survey of Expectations."
37. Sparrow and Sparrow, "In the Hands of Machines?"
38. Friedman, "Value-Sensitive Design"; Friedman and Hendry, *Value Sensitive Design*.
39. van Wynsberghe, "Designing Robots."
40. von Schomberg, *Towards Responsible Research*.
41. Stahl and Coeckelbergh, "Ethics."

Chapter 5

1. http://moralmachine.mit.edu/.
2. Awad et al., "Moral Machine."
3. Nyholm and Smids, "Ethics of Accident-Algorithms."
4. Levin and Wong, "Self-Driving Uber."
5. Anderson and Anderson, *Machine Ethics*, 9.
6. Winfield, "Making an Ethical Machine."
7. Asimov, "Runaround."

8. Winfield, Blum, and Liu, "Towards an Ethical Robot."
9. Wallach and Allen, *Moral Machines*, 7.
10. Johnson, "Computer Systems."
11. Coeckelbergh, "Moral Appearances."
12. Clark, "Asimov's Laws."
13. Bostrom, *Superintelligence*, 139.
14. Wallach and Allen, *Moral Machines*, 8, 9, 26.
15. Floridi and Sanders, "On the Morality."
16. Sullins, "When Is a Robot a Moral Agent?," 29.
17. Nyholm, *Humans and Robots*, 55.
18. Helveke and Nida-Rümelin, "Responsibility for Crashes."
19. Matthias, "Responsibility Gap."
20. Fischer and Ravizza, *Responsibility and Control*; Rudy-Hiller, "Epistemic Condition."
21. Aristotle, *Complete Works*, 1109b30–1111b5.
22. van de Poel et al., "Many Hands."
23. Fischer and Ravizza, *Responsibility and Control*, 13.
24. Aristotle, *Complete Works*, 1111a3–1111a5.
25. Rudy-Hiller, "Epistemic Condition."
26. Duff, "Who Is Responsible?"
27. Smith, "Responsibility as Answerability."
28. Coeckelbergh, "AI, Responsibility Attribution."
29. Coeckelbergh, "AI, Responsibility Attribution."
30. Loh and Loh, "Autonomy and Responsibility," 36.
31. Coeckelbergh, "Virtual Moral Agency."
32. Sullins, "When Is a Robot a Moral Agent?"

Chapter 6

1. Parke, "Is it Cruel?"
2. Leopold, "HitchBOT."
3. Wakefield, "Can You Murder."
4. Suzuki et al., "Measuring Empathy."
5. Darling, "Who's Johnny," 181.
6. Horstmann et al., "Do a Robot's Social Skills and Its Objection Discourage Interactants?," 20–21.
7. Ishiguro, "Android Science."
8. Mori, "Uncanny Valley"; MacDorman and Ishiguro, "Uncanny Advantage."
9. Tognazzini, "Principles, Techniques, and Ethics of Stage Magic."
10. Flusser, *Shape of Things*, 17.

11. Coeckelbergh, "How to Describe."
12. Bryson, "Robots Should Be Slaves."
13. Gunkel, "Other Question."
14. Kant, *Lectures on Ethics.*
15. Darling, "Extending Legal Protection."
16. Coeckelbergh, "You, Robot."
17. Coeckelbergh, *Growing Moral Relations*; Coeckelbergh, "Moral Standing of Machines."
18. Coeckelbergh, *Growing Moral Relations.*
19. Gunkel, *Machine Question*; Gunkel, "Other Question."
20. See, for example, Danaher, "Welcoming Robots"; Coeckelbergh, "Should We Treat Teddy Bear 2.0 as a Kantian Dog?"
21. Coeckelbergh, "Why Care about Robots?"
22. Coeckelbergh, *Moved by Machines.*
23. Coeckelbergh, "How to Describe"; Zimmerli, "Deus Malignus."
24. Danaher, *Automation and Utopia*, 186.
25. Carpenter, "Deus Sex Machina."
26. Coeckelbergh, "How to Use"; Coeckelbergh, "Should We Treat Teddy Bear 2.0 as a Kantian Dog?"

Chapter 7

1. Note that the term "unmanned" is problematic in terms of gender.
2. Piper, "Death by Algorithm."
3. https://www.stopkillerrobots.org/.
4. McKay, "19 Climate Change Activists."
5. Lin, Bekey, and Abney, *Autonomous Military Robotics.*
6. Kreps, "Ground the Drones?"
7. Arkin, "Case for Ethical Autonomy."
8. Asaro, "What Should We Want from a Robot Ethic?"
9. Müller, "Autonomous Killer Robots," 76.
10. Sullins, "RoboWarfare."
11. Byrne, "Making Drones."
12. Sparrow, "Killer Robots," 70, 71, 75.
13. Müller, "Autonomous Killer Robots," 75–76, 78.
14. See, for example, Article 36, "Killing by Machine."
15. Santoni de Sio and van den Hoven, "Meaningful Human Control."
16. See, for example, Gross, "Assassination."
17. Heyns, *Report*, 20–21.
18. Grossman, *On Killing.*

19. Sharkey, "Killing Made Easy."
20. Coeckelbergh, "Drones, Information Technology, and Distance."
21. Linebaugh, "I Worked on the U.S. Drone Program."
22. Bumiller, "Day Job"; Coeckelbergh, "Drones, Information Technology, and Distance"; Coeckelbergh, "Drones, Morality, and Vulnerability."
23. Williams, "Distant Intimacy"; Coeckelbergh, "Drones, Information Technology, and Distance."
24. McCammon, "Warfare."
25. Press, "Wounds."
26. Monbiot, "With Its Deadly Drones."
27. Enemark, *Armed Drones*, 7, 4.
28. Coeckelbergh, "Drones, Morality, and Vulnerability."
29. Heyns, *Report*, 17.
30. Møller, "Secretary-General's Message."
31. Campaign to Stop Killer Robots, "Parliamentary Actions in Europe."

Chapter 8

1. https://conferences.au.dk/robo-philosophy-2020-at-aarhus-university/previous-conferences/rp2018/.
2. Funk, Seibt, and Coeckelbergh, "Why Do/Should We Build Robots?"
3. Descartes, *Discourse on Method*, part 5.
4. Coeckelbergh, *Human Being @ Risk*; Coeckelbergh, "Robotic Appearances."
5. Moravec, *Mind Children*, 1.
6. Kurzweil, *Singularity Is Near*.
7. Bostrom, *Superintelligence*.
8. Tegmark, *Life 3.0*, 335.
9. Haraway, "Cyborg Manifesto," 295, 292.
10. Latour, *We Have Never Been Modern*.
11. Braidotti, *Posthuman*, 3, 43, 94, 90, 92, 102.
12. van Wynsberghe and Donhauser, "Dawning."

BIBLIOGRAPHY

Abney, Keith. "Robotics, Ethical Theory, and Metaethics: A Guide for the Perplexed." In *Robot Ethics: The Ethical and Social Implications of Robotics*, edited by Patrick Lin, George Bekey, and Keith Abney, 35–52. Cambridge, MA: MIT Press, 2012.

Anderson, Michael, and Susan Anderson, eds. *Machine Ethics*. Cambridge: Cambridge University Press, 2011.

Archer, Joseph. "'Friendly' Hospital Robot Begins Trials to Help Stressed Nurses." *Telegraph*, September 19, 2018. https://www.telegraph.co.uk/technology/2018/09/19/friendly-hospital-robot-begins-trials-help-stressed-nurses/.

Aristotle. *The Complete Works of Aristotle*. Edited by Jonathan Barnes. 2 vols. Princeton, NJ: Princeton University Press, 1984.

Arkin, Ronald. "The Case for Ethical Autonomy in Unmanned Systems." *Journal of Military Ethics* 9, no. 4 (2010): 332–341.

Article 36. "Killing by Machine: Key Issues for Understanding Meaningful Human Control." April 2015. https://article36.org/wp-content/uploads/2013/06/KILLING_BY_MACHINE_6.4.15.pdf.

Asaro, Peter. "How Just Could a Robot War Be?" In *Current Issues in Computing and Philosophy*, edited by Philip Brey, Adam Briggle, and Katinka Waelbers, 50–64. Amsterdam: IOS Press, 2008.

Asaro, Peter. "What Should We Want from a Robot Ethic?" *International Review of Information Ethics* 6 (2006): 9–16.

Asimov, Isaac. "Runaround: A Short Story." *Astounding Science-Fiction* 29, no. 1 (March 1942): 94–103.

Awad, Edmond, Sohan Dsouza, Richard Kim, Jonathan Schulz, Joseph Henrich, Azim Shariff, Jean-François Bonnefon, and Iyad Rahwan. "The Moral Machine Experiment." *Nature* 563 (2018): 59–64.

Bartneck, Christoph, Christoph Lütge, Alan R. Wagner, and Sean Welsh. *An Introduction to Ethics in Robotics and AI*. Cham, Switzerland: Springer, 2021.

Beauchamp, Tom, and James Childress. *Principles of Biomedical Ethics*. Oxford: Oxford University Press, 1994.

Benjamin, Ruha. *Race after Technology: Abolitionist Tools for the New Jim Code*. Cambridge, UK: Polity Press, 2019.

Boddington, Paula. *Towards a Code of Ethics for Artificial Intelligence*. Cham, Switzerland: Springer, 2017.

Borenstein, Jason, and Yvette Pearson. "Robot Caregivers: Harbingers of Expanded Freedom for All?" *Ethics and Information Technology* 12 (2010): 277–288.

Bostrom, Nick. *Superintelligence: Paths, Dangers, Strategies*. Oxford: Oxford University Press, 2014.

Botting, Eileen Hunt. *Mary Shelley and the Rights of the Child*. Philadelphia: University of Pennsylvania Press, 2018.

Braidotti, Rosi. *The Posthuman*. Cambridge, UK: Polity Press, 2013.

Brynjolfsson, Erik, and Andrew McAfee. *The Second Machine Age: Work, Progress, and Prosperity in a Time of Brilliant Technologies*. New York: W. W. Norton and Company, 2014.

Bryson, Joanna. "Robots Should Be Slaves." In *Close Engagements with Artificial Companions: Key Social, Psychological, Ethical and Design Issues*, edited by Yorick Wilks, 63–74. Amsterdam: John Benjamins, 2010.

Bumiller, Elisabeth. "A Day Job Waiting for a Kill Shot a World Away." *New York Times*, July 30, 2012. https://www.nytimes.com/2012/07/30/us/drone-pilots-waiting-for-a-kill-shot-7000-miles-away.html.

Byrne, Edmund. "Making Drones to Kill Civilians: Is It Ethical?" *Journal of Business Ethics* 147 (2015): 81–93.

Calo, M. Ryan. "Robots and Privacy." In *Robot Ethics: The Ethical and Social Implications of Robotics*, edited by Patrick Lin, George Bekey, and Keith Abney, 187–201. Cambridge, MA: MIT Press, 2012.

Campaign to Stop Killer Robots. "Parliamentary Actions in Europe." July 10, 2018. https://www.stopkillerrobots.org/2018/07/parliaments-2/.

Carpenter, Julie. "Deus Sex Machina: Loving Robot Sex Workers and the Allure of an Insincere Kiss." In *Robot Sex: Social and Ethical Implications*, edited by John Danaher and Neil McArthur, 261–287. Cambridge, MA: MIT Press, 2017.

Carpenter, Julie. "The Existential Robot: Living with Robots May Teach Us to Be Better Humans." *Issues* 108 (2014): 39–42.

Carpenter, Julie, Joan M. Davis, Norah Erwin Stewart, Tiffany R. Lee, John D. Bransford, and Nancy Vye. "Gender Representation and Humanoid Robots Designed for Domestic Use." *International Journal of Social Robotics* 1 (2009): 261–265.

Clark, Roger. "Asimov's Laws of Robotics: Implications for Information Technology." In *Machine Ethics*, edited by Michael Anderson and Susan Anderson, 254–284. Cambridge: Cambridge University Press, 2011.

Coeckelbergh, Mark. *AI Ethics*. Cambridge, MA: MIT Press, 2020.

Coeckelbergh, Mark. "AI, Responsibility Attribution, and a Relational Justification of Explainability." *Science and Engineering Ethics* 26, no. 4 (2020): 2051–2068.

Coeckelbergh, Mark. "Artificial Agents, Good Care, and Modernity." *Theoretical Medicine and Bioethics* 36 (2015): 265–277.

Coeckelbergh, Mark. "Care Robots and the Future of ICT-Mediated Elderly Care: A Response to Doom Scenarios." *AI and Society* 31, no. 4 (2016): 455–462.

Coeckelbergh, Mark. "Care Robots, Virtual Virtue, and the Best Possible Life." In *The Good Life in a Technological Age*, edited by Philip Brey, Adam Briggle, and Ed Spence, 281–292. New York: Routledge, 2012.

Coeckelbergh, Mark. "Drones, Information Technology, and Distance: Mapping the Moral Epistemology of Remote Fighting." *Ethics and Information Technology* 15, no. 2 (2013): 87–98.

Coeckelbergh, Mark. "Drones, Morality, and Vulnerability: Two Arguments against Automated Killing." In *The Future of Drone Use: Technologies, Opportunities and Privacy Issues*, edited by Bart Custers, 229–237. The Hauge: T. M. C. Asser Press, 2016.

Coeckelbergh, Mark. "E-Care as Craftsmanship: Virtuous Work, Skilled Engagement, and Information Technology in Health Care." *Medicine, Healthcare and Philosophy* 16, no. 4 (2013): 807–816.

Coeckelbergh, Mark. *Growing Moral Relations: Critique of Moral Status Ascription*. New York: Palgrave Macmillan, 2012.

Coeckelbergh, Mark. "Health Care, Capabilities, and AI Assistive Technologies." *Ethical Theory and Moral Practice* 13 (2010): 181–190.

Coeckelbergh, Mark. "How I Learned to Love the Robot: Capabilities, Information Technologies, and Elderly Care." In *The Capability Approach, Technology and Design*, edited by Ilse Oosterlaken and Jeroen van den Hoven, 77–86. Dordrecht: Springer, 2012.

Coeckelbergh, Mark. "How to Describe and Evaluate 'Deception' Phenomena: Recasting the Metaphysics, Ethics, and Politics of ICTs in Terms of Magic and Performance and Taking a Relational and Narrative Turn." *Ethics and Information Technology* 20, no. 2 (2018): 71–85.

Coeckelbergh, Mark. "How to Use Virtue Ethics for Thinking about the Moral Standing of Social Robots: A Relational Interpretation in Terms of Practices, Habits, and Performance." *International Journal of Social Robotics* 13, no. 1 (2021): 31–40.

Coeckelbergh, Mark. *Human Being @ Risk: Enhancement, Technology, and the Evaluation of Vulnerability Transformations*. New York: Springer, 2013.

Coeckelbergh, Mark. "Moral Appearances: Emotions, Robots, and Human Morality." *Ethics and Information Technology* 12, no. 3 (2010): 235–241.

Coeckelbergh, Mark. "The Moral Standing of Machines: Towards a Relational and Non-Cartesian Moral Hermeneutics." *Philosophy and Technology* 27, no. 1 (2014): 61–77.

Coeckelbergh, Mark. *Moved by Machines: Performance Metaphors and Philosophy of Technology*. New York: Routledge, 2019.

Coeckelbergh, Mark. *New Romantic Cyborgs*. Cambridge, MA: MIT Press, 2017.

Coeckelbergh, Mark. "Robotic Appearances and Forms of Life: A Phenomenological-Hermeneutical Approach to the Relation between Robotics and Culture." In *Robotics in Germany and Japan: Philosophical and Technical Perspectives*, edited by Michael Funk and Bernhard Irrgang, 59–68. Berlin: Peter Lang, 2014.

Coeckelbergh, Mark. "Should We Treat Teddy Bear 2.0 as a Kantian Dog? Four Arguments for the Indirect Moral Standing of Personal Social Robots, with Implications for Thinking about Animals and Humans." *Minds and Machines* (December 30, 2020). https://doi.org/10.1007/s11023-020-09554-3.

Coeckelbergh, Mark. "Technology Games/Gender Games: From Wittgenstein's Toolbox and Language Games to Gendered Robots and Biased Artificial Intelligence." In *Feminist Philosophy of Technology*, edited by Janina Loh and Mark Coeckelbergh, 27–38. Stuttgart: J. B. Metzler, 2020.

Coeckelbergh, Mark. "Virtual Moral Agency, Virtual Moral Responsibility." *AI and Society* 24, no. 2 (2009): 181–189.

Coeckelbergh, Mark. "Why Care about Robots? Empathy, Moral Standing, and the Language of Suffering." *Kairos: Journal of Philosophy and Science* 20, no. 1 (2018): 141–158.

Coeckelbergh, Mark. "You, Robot: On the Linguistic Construction of Artificial Others." *AI and Society* 26, no. 1 (2011): 61–69.

Coeckelbergh, Mark, Janina Loh, Michael Funk, Johanna Seibt, and Marco Nørskov, eds. *Envisioning Robots in Society: Power, Politics, and Public Space*. Proceedings of Robophilosophy 2018/TRANSOR 2018, University of Vienna, February 2018. Amsterdam: IOC Press, 2018.

Coeckelbergh, Mark, Christina Pop, Ramona Simut, Andreea Peca, Sebastian Pintea, Daniel David, and Bram Vanderborght. "A Survey of Expectations about the Role of Robots in Robot-Assisted Therapy for Children with ASD: Ethical Acceptability, Trust, Sociability, Appearance, and Attachment." *Science and Engineering Ethics* 22, no. 1 (2016): 47–65.

Danaher, John. *Automation and Utopia: Human Flourishing in a World without Work*. Cambridge, MA: Harvard University Press, 2019.

Danaher, John. "Welcoming Robots into the Moral Circle: A Defense of Ethical Behaviourism." *Science and Engineering Ethics* 26, no. 4 (2019): 2023–2049. https://doi.org/10.1007/s11948-019-00119-x.

Danaher, John, and Neil McArthur. *Robot Sex: Social and Ethical Implications*. Cambridge, MA: MIT Press, 2017.

Darling, Kate. "Extending Legal Protection to Social Robots." *IEEE Spectrum*, September 10, 2012. http://spectrum.ieee.org/automaton/robotics/artificial-intelligence/extending-legal-protection-to-social-robots.

Darling, Kate. "'Who's Johnny?' Anthropomorphic Framing in Human-Robot Interaction, Integration, and Policy." In *Robot Ethics 2.0: From Autonomous Cars to Artificial Intelligence*, edited by Patrick Lin, Keith Abney, and Ryan Jenkins, 173–188. New York: Oxford University Press, 2017.

Descartes, René. *Discourse on Method*. In *Discourse on Method and Meditations*. Translated by Laurence J. Lafleur. Indianapolis, IN: Bobbs-Merrill, 1960.

Dreyfus, Hubert L., and Stuart E. Dreyfus. "Towards a Phenomenology of Ethical Expertise." *Human Studies* 14, no. 4 (1991): 229–250.

Dreyfus, Stuart E., and Hubert L. Dreyfus. *A Five-Stage Model of the Mental Activities Involved in Direct Skill Acquisition*. Berkeley: Operations Research Center, University of California, 1980.

Duff, R. Antony. "Who Is Responsible, for What, to Whom?" *Ohio State Journal of Criminal Law* 2 (2005): 441–461.

Enemark, Christian. *Armed Drones and the Ethics of War: Military Virtue in a Post-Heroic Age*. Abingdon, UK: Routledge, 2016.

Fischer, John M., and Mark Ravizza. *Responsibility and Control: A Theory of Moral Responsibility*. Cambridge: Cambridge University Press, 1998.

Fletcher, Sarah R., and Philip Webb. "Industrial Robot Ethics: The Challenges of Closer Human Collaboration in Future Manufacturing Systems." In *A World with Robots: International Conference on Robot Ethics: ICRE 2015*, edited by Maria Isabel Aldinhas Ferreira, Joao Silva Sequeira, Mohammad O. Tokhi, Endre E. Kadar, and Gurvinder S. Virk, 159–169. Cham, Switzerland: Springer, 2017.

Floridi, Luciano, and J. W. Sanders. "On the Morality of Artificial Agents." *Minds and Machines* 14, no. 3 (2004): 349–379.

Flusser, Vilém. *Shape of Things: A Philosophy of Design*. London: Reaction Books, 1999.

Ford, Martin. *Rise of the Robots: Technology and the Threat of Jobless Future*. New York: Basic Books, 2015.

Fosch-Villaronga, Eduard. *Robots, Healthcare, and the Law: Regulation Automation in Personal Care*. Abingdon, UK: Routledge, 2020.

Foster, Malcolm. "Aging Japan: Robots May Have Role in Future of Elder Care." *Reuters*, March 28, 2018. https://www.reuters.com/article/us-japan-ageing-robots-widerimage/aging-japan-robots-may-have-role-in-future-of-elder-care-idUSKBN1H33AB.

Frey, Carl Benedikt, and Michael A. Osborne. "The Future of Employment: How Susceptible Are Jobs to Computerisation?" Oxford Martin School Working Paper No. 7, 2013. https://www.oxfordmartin.ox.ac.uk/downloads/academic/future-of-employment.pdf.

Friedman, Batya. "Value-Sensitive Design." *Interactions* 3, no. 6 (1996): 16–23.

Friedman, Batya, and David G. Hendry. *Value Sensitive Design: Shaping Technology with Moral Imagination*. Cambridge, MA: MIT Press, 2019.

Funk, Michael, Johanna Seibt, and Mark Coeckelbergh. "Why Do/Should We Build Robots? Summary of a Plenary Discussion Session." In *Envisioning Robots in Society: Power, Politics, and Public Space*, edited by Mark Coeckelbergh, Janina Loh, Michael Funk, Johanna Seibt, and Marco Nørskov, 369–384. Proceedings of Robophilosophy 2018/TRANSOR 2018, University of Vienna, February 2018. Amsterdam: IOC Press, 2018.

Gross, Michael L. "Assassination and Targeted Killing: Law Enforcement, Execution or Self-Defence?" *Journal of Applied Philosophy* 23, no. 3 (2006): 323–334.

Grossman, Dave. *On Killing: The Psychological Cost of Learning to Kill in War and Society*. New York: Little, Brown and Company, 2009.

Gunkel, David. *An Introduction to Communication and Artificial Intelligence*. Cambridge, UK: Polity Press, 2020.

Gunkel, David. *The Machine Question: Critical Perspectives on AI, Robots, and Ethics*. Cambridge, MA: MIT Press, 2012.

Gunkel, David. "The Other Question: Can and Should Robots Have Rights?" *Ethics and Information Technology* 20, no. 2 (2018): 87–99.

Haraway, Donna. "A Cyborg Manifesto." In *The Cybercultures Reader*, edited by David Bell and Barbara M. Kennedy, 291–324. London: Routledge, 2000.

Heidegger, Martin. *The Question concerning Technology and Other Essays*. Translated by William Lovitt. New York: Harper and Row, 1977.

Helveke, Alexander, and Julian Nida-Rümelin. "Responsibility for Crashes of Autonomous Vehicles: An Ethical Analysis." *Science and Engineering Ethics* 21, no. 3 (2015): 619–630.

Heyns, Christof. *Report of the Special Rapporteur on Extrajudicial, Summary or Arbitrary Executions*. United Nations, Human Rights Council, April 9, 2013 (A/HRC/23/47). http://www.ohchr.org/Documents/HRBodies/HRCouncil/RegularSession/Session23/A-HRC-23-47_en.pdf.

Hildebrandt, Mireille. *Smart Technologies and the End(s) of Law: Novel Entanglements of Law and Technology*. Cheltenham, UK: Elgar, 2015.

Horstmann, Aike C., Nikolai Bock, Eva Linhuber, Jessica M. Szczuka, Carolin Straßmann, and Nicole C. Krämer. "Do a Robot's Social Skills and Its Objection

Discourage Interactants from Switching the Robot Off?" *PLoS ONE* 13, no. 7 (2018): e0201581. https://doi.org/10.1371/journal.pone.0201581.

Isaac, Alistair M. C., and Will Bridewell. "White Lies on Silver Tongues: Why Robots Need to Deceive (and How)." In *Robot Ethics 2.0: From Autonomous Cars to Artificial Intelligence*, edited by Patrick Lin, Keith Abney, and Ryan Jenkins, 157–172. New York: Oxford University Press, 2017.

Ishiguro, Hiroshi. 2006. "Android Science: Toward a New Cross-Interdisciplinary Framework." *Scientific American* 294, no. 5 (2006): 32–34.

Johnson, Deborah. "Computer Systems: Moral Entities but Not Moral Agents." *Ethics and Information Technology* 8, no. 4 (2006): 195–204.

Kant, Immanuel. *Lectures on Ethics*. Edited by Peter Heath and J. B. Schneewind. Translated by Peter Heath. Cambridge: Cambridge University Press, 1997.

Kitwood, Tom. *Dementia Reconsidered: The Person Comes First*. Buckingham, UK: Open University Press, 1997.

Kreps, Sarah. "Ground the Drones? The Real Problem with Unmanned Aircraft." *Foreign Affairs*, December 4, 2013. https://www.foreignaffairs.com/articles/2013-12-04/ground-drones.

Kurzweil, Ray. *The Singularity Is Near: When Humans Transcend Biology*. New York: Penguin, 2005.

Latour, Bruno. *We Have Never Been Modern*. Translated by Catherine Porter. Cambridge, MA: Harvard University Press, 1993.

Leopold, Todd. "HitchBOT, the Hitchhiking Robot, Gets Beheaded in Philadelphia." CNN, August 4, 2015. https://edition.cnn.com/2015/08/03/us/hitchbot-robot-beheaded-philadelphia-feat/index.html.

Levin, Sam, and Julia Carrie Wong. "Self-Driving Uber Kills Arizona Woman in First Fatal Crash Involving Pedestrian." *Guardian*, March 19, 2018. https://www.theguardian.com/technology/2018/mar/19/uber-self-driving-car-kills-woman-arizona-tempe.

Liao, S. Matthew, ed. *Ethics of Artificial Intelligence*. New York: Oxford University Press, 2020.

Lin, Patrick, Keith Abney, and George Bekey. "Robot Ethics: Mapping the Issues for a Mechanized World." *Artificial Intelligence* 175 (2011): 942–949.

Lin, Patrick, Keith Abney, and Ryan Jenkins, eds. *Robot Ethics 2.0: From Autonomous Cars to Artificial Intelligence*. New York: Oxford University Press, 2017.

Lin, Patrick, George Bekey, and Keith Abney. *Autonomous Military Robotics: Risk, Ethics, and Design*. Report commissioned under the US Department of the Navy, Office of Naval Research, award #N00014-07-1-1152. San Luis Obispo: California Polytechnic State University, 2008.

Lin, Patrick, George Bekey, and Keith Abney, eds. *Robot Ethics: The Ethical and Social Implications of Robotics*. Cambridge, MA: MIT Press, 2012.

Linebaugh, Heather. "I Worked on the U.S. Drone Program: The Public Should Know What Really Goes On." *Guardian*, December 29, 2013. https://www.theguardian.com/commentisfree/2013/dec/29/drones-us-military.

Loh, Wulf, and Janina Loh. "Autonomy and Responsibility in Hybrid Systems: The Example of Autonomous Cars." In *Robot Ethics 2.0: From Autonomous Cars to Artificial Intelligence*, edited by Patrick Lin, Keith Abney, and Ryan Jenkins, 35–50. New York: Oxford University Press, 2017.

MacDorman, Karl F., and Hiroshi Ishiguro. "The Uncanny Advantage of Using Androids in Cognitive and Social Science Research." *Interaction Studies* 7, no. 3 (2006): 297–337.

Marx, Karl. *Capital: A Critique of Political Economy*. Translated by Ben Fowkes. London: Penguin, 1990.

Matsuzaki, Hironori, and Gesa Lindemann. "The Autonomy-Safety-Paradox of Service Robotics in Europe and Japan: A Comparative Analysis." *AI and Society* 31, no. 4 (2016): 501–517.

Matthias, Andreas. "The Responsibility Gap: Ascribing Responsibility for the Actions of Learning Automata." *Ethics and Information Technology* 6, no. 3 (2004): 175–183.

Mavroforou, Anna, Emmanuel Michalodimitrakis, Konstantinos Hatzitheo-Filou, and Athanasios Giannoukas. "Legal and Ethical Issues in Robotic Surgery." *International Union of Angiology* 29, no. 1 (2010): 75–79.

McCammon, Sarah. "The Warfare May Be Remote but the Trauma Is Real." NPR, April 24, 2017. https://www.npr.org/2017/04/24/525413427/for-drone-pilots-warfare-may-be-remote-but-the-trauma-is-real?t=1571811594970&t=1614700024482.

McKay, Tom. "19 Climate Change Activists Arrested for Drone Protest against Heathrow Airport Expansion." *Gizmodo*, September 14, 2019. https://earther.gizmodo.com/19-extinction-rebellion-activists-arrested-in-drone-pro-1838122386.

McKinsey Global Institute. *Jobs Lost, Jobs Gained: Workforce Transitions in a Time of Automation*. New York: McKinsey Global Institute, December 2017.

Møller, Michael. "Secretary-General's Message to Meeting of the Group of Governmental Experts on Emerging Technologies in the Area of Lethal Autonomous Weapons Systems." Talk presented at the United Nations Office, Geneva, March 25, 2019. https://www.un.org/sg/en/content/sg/statement/2019-03-25/secretary-generals-message-meeting-of-the-group-of-governmental-experts-emerging-technologies-the-area-of-lethal-autonomous-weapons-systems.

Monbiot, George. "With Its Deadly Drones, the U. S. Is Fighting a Coward's War." *Guardian*, January 30, 2012. https://www.theguardian.com/commentisfree/2012/jan/30/deadly-drones-us-cowards-war.

Moravec, Hans. *Mind Children: The Future of Robot and Human Intelligence*. Cambridge, MA: Harvard University Press, 1988.

Mori, Masahiro. "The Uncanny Valley." Translated by Karl F. MacDorman and Norri Kageki. *IEEE Robotics and Automation Magazine* 19, no. 2 (2012): 98–100.

Müller, Vincent. "Autonomous Killer Robots Are Probably Good News." In *Drones and Responsibility: Legal, Philosophical and Sociotechnical Perspectives on Remotely Controlled Weapons*, edited by Ezio Di Nucci and Filippo Santoni de Sio, 67–81. London: Routledge, 2016.

Nordrum, Amy. "Cyderdyne's HAL Exoskeleton Helps Patients Walk Again in First Treatments at U.S. Facility." *IEEE Spectrum*, January 3, 2019. https://spectrum.ieee.org/the-human-os/biomedical/bionics/cyberdynes-hal-medical-exoskeleton-helps-patients-walk-again-at-first-us-facility.

Nozick, Robert. *Anarchy, State, and Utopia*. New York: Basic Books, 1974.

Nussbaum, Martha. *Frontiers of Justice*. Cambridge, MA: Belknap Press, 2006.

Nussbaum, Martha. *Women and Human Development: The Capabilities Approach*. Cambridge: Cambridge University Press, 2000.

Nussbaum, Martha, and Amartya Sen. *The Quality of Life*. Oxford: Clarendon Press, 1993.

Nyholm, Sven. *Humans and Robots: Ethics, Agency, and Anthropocentrism*. London: Rowman and Littlefield, 2020.

Nyholm, Sven, and Jilles Smids. 2016. "The Ethics of Accident-Algorithms for Self-Driving Cars: An Applied Trolley Problem?" *Ethical Theory and Moral Practice* 19, no. 5 (2016): 1275–1289.

Parke, Phoebe. "Is It Cruel to Kick a Robot Dog?" CNN, February 13, 2015. https://edition.cnn.com/2015/02/13/tech/spot-robot-dog-google/index.html.

Piper, Kelsey. "Death by Algorithm: The Age of Killer Robots Is Closer than You Think." *Vox*, June 21, 2019. https://www.vox.com/2019/6/21/18691459/killer-robots-lethal-autonomous-weapons-ai-war.

Press, Eyal. "The Wounds of the Drone Warrior." *New York Times Magazine*, June 13, 2018. https://www.nytimes.com/2018/06/13/magazine/veterans-ptsd-drone-warrior-wounds.html.

PricewaterhouseCoopers. *Fourth Industrial Revolution for the Earth: Harnessing Artificial Intelligence for the Earth*. PwC Network, 2018.

PricewaterhouseCoopers. *Will Robots Really Steal Our Jobs? An International Analysis of the Potential Long Term Impact of Automation*. PwC UK and PwC Network, 2018.

Reese, Hope. "Wi-Fi-Enabled 'Hello Barbie' Records Conversations with Kids and Uses AI to Talk Back." *TechRepublic*, November 10, 2015. https://www.techrepublic.com/article/wi-fi-enabled-hello-barbie-records-conversations-with-kids-and-uses-ai-to-talk-back/.

Richardson, Kathleen. "The Asymmetrical 'Relationship': Parallels between Prostitution and the Development of Sex Robots." *ACM SIGCAS Computers and Society—Special Issue on Ethicomp* 45, no. 3 (2016): 290–293.

Rohrlich, Justin. "Why a U.S. Soldier Turned against Drone Warfare." *Quartz*, October 16, 2019. https://qz.com/1725819/why-a-us-soldier-turned-against-drone-warfare/.

Rudy-Hiller, Fernando. "The Epistemic Condition for Moral Responsibility." *Stanford Encyclopedia of Philosophy*, edited by Edward N. Zalta. Fall 2018 ed. https://plato.stanford.edu/entries/moral-responsibility-epistemic/.

Santoni de Sio, Filippo, and Jeroen van den Hoven. "Meaningful Human Control over Autonomous Systems: A Philosophical Account." *Frontiers in Robotics and AI* (February 28, 2018). https://www.frontiersin.org/articles/10.3389/frobt.2018.00015/full#B5.

Schwab, Klaus. *The Fourth Industrial Revolution*. New York: Crown Publishing Group, 2016.

Sennett, Richard. *The Craftsman*. New Haven, CT: Yale University Press, 2008.

Servoz, Michel. *The Future of Work? Work of the Future! On How Artificial Intelligence, Robotics and Automation Are Transforming Jobs and the Economy in Europe*. Brussels: European Commission, 2019.

Sharkey, Amanda, and Noel Sharkey. "Granny and the Robots: Ethical Issues in Robot Care for the Elderly." *Ethics and Information Technology* 14 (2012): 27–40.

Sharkey, Noel. 2012. "Killing Made Easy: From Joysticks to Politics." In *Robot Ethics: The Ethical and Social Implications of Robotics*, edited by Patrick Lin, George Bekey, and Keith Abney, 111–128. Cambridge, MA: MIT Press, 2012.

Sharkey, Noel, and Amanda Sharkey. "The Crying Shame of Robot Nannies: An Ethical Appraisal." *Interaction Studies* 11, no. 2 (2010): 161–190.

Smith, Angela M. "Responsibility as Answerability." *Inquiry* 58, no. 2 (2015): 99–126.

Smith, Marquard. *The Erotic Doll: A Modern Fetish*. New Haven, CT: Yale University Press, 2013.

Sorell, Tom, and Heather Draper. "Robot Carers, Ethics, and Older People." *Ethics and Information Technology* 16 (2014): 183–195.

Sparrow, Robert. "Killer Robots." *Journal of Applied Philosophy* 24, no. 1 (2007): 62–77.

Sparrow, Robert. "Robots in Aged Care: A Dystopian Future?" *AI and Society* 31, no. 4 (2016): 445–454.

Sparrow, Robert, and Linda Sparrow. "In the Hands of Machines? The Future of Aged Care." *Minds and Machines* 16, no. 2 (2006): 141–161.

Stahl, Bernd C., and Mark Coeckelbergh. "Ethics of Healthcare Robotics: Towards Responsible Research and Innovation." *Robotics and Autonomous Systems* 86 (2016): 152–161.

Sullins, John. "Robots, Love, and Sex: The Ethics of Building a Love Machine." *IEEE Transactions on Affective Computing* 3, no. 4 (2012): 398–409.

Sullins, John. "RoboWarfare: Can Robots Be More Ethical than Humans on the Battlefield?" *Ethics and Information Technology* 12, no. 3 (2010): 263–275.

Sullins, John. "When Is a Robot a Moral Agent?" *International Review of Information Ethics* 6 (2006): 23–30.

Suzuki, Yutaka, Lisa Galli, Ayaka Ikeda, Shoji Itakura, and Michiteru Kitazaki. "Measuring Empathy for Human and Robot Hand Pain Using Electroencephalography." *Scientific Reports* 5 (2015), article no. 15924. https://doi.org/10.1038/srep15924.

Tegmark, Max. *Life 3.0: Being Human in the Age of Artificial Intelligence*. London: Allen Lane, 2017.

Tognazzini, Bruce. "Principles, Techniques, and Ethics of Stage Magic and Their Application to Human Interface Design." *CHI '93: Proceedings of the INTERACT '93 and CHI '93 Conference on Human Factors in Computing Systems*, 355–362. New York: Association for Computing Machinery, 1993. http://dl.acm.org/citation.cfm?id=169284.

Turkle, Sherry. *Reclaiming Conversation: The Power of Talk in a Digital Age*. New York: Penguin Books, 2015.

Turner, Jacob. *Robot Rules: Regulating Artificial Intelligence*. Cham, Switzerland: Palgrave Macmillan, 2019.

UNI Global Union. *Top 10 Principles for Workers' Data Privacy and Protection*. Nyon, Switzerland: UNI Global Union. http://www.thefutureworldofwork.org/media/35421/uni_workers_data_protection.pdf.

van de Poel, Ibo, Jessica Nihlén Fahlquist, Neelke Doorn, Sjoerd D. Zwart, and Lambèr Royakkers. "The Problem of Many Hands: Climate Change as an Example." *Science and Engineering Ethics* 18, no. 1 (2012): 49–67.

van den Hoven, Jeroen, Martijn Blaauw, Wolte Pieters, and Martijn Warnier. "Privacy and Information Technology." *Stanford Encyclopedia of Philosophy*, edited by Edward N. Zalta. Winter 2019 ed. https://plato.stanford.edu/entries/it-privacy/.

van Wynsberghe, Aimee. "Designing Robots for Care: Care Centered Value-Sensitive Design." *Science and Engineering Ethics* 19, no. 2 (2013): 407–433.

van Wynsberghe, Aimee, and Justin Donhauser. "The Dawning of the Ethics of Environmental Robots." *Science and Engineering Ethics* 24, no. 6 (2018): 1777–1800.

van Wynsberghe, Aimee, and Shuhong Li. "A Paradigm Shift for Robot Ethics: From HRI to Human-Robot-System Interaction (HRSI)." *Medicolegal and Bioethics* 9 (2019): 11–21.

Veal, Anthony J. "The Leisure Society I: Myths and Misperceptions, 1960–79." *World Leisure Journal* 53, no. 3 (2011): 203–227.

Véliz, Carissa. *Privacy Is Power: Why and How You Should Take Back Control of Your Data*. London: Penguin, 2020.

von Schomberg, Rene, ed. *Towards Responsible Research and Innovation in the Information and Communication Technologies and Security Technologies Fields*. Luxembourg: Publication Office of the European Union, 2011. http://ec.europa.eu/research/science-society/document_library/pdf_06/mep-rapport-2011_en.pdf.

Wakefield, Jane. "Can You Murder a Robot?" BBC, March 17, 2019. https://www.bbc.com/news/technology-47090174.

Wakefield, Jane. "Foxconn Replaces '60,000 Factory Workers with Robots.'" BBC, Technology, May 25, 2016. https://www.bbc.com/news/technology-36376966.

Wallach, Wendell, and Colin Allen. *Moral Machines: Teaching Robots Right from Wrong*. Oxford: Oxford University Press, 2009.

Wiener, Norbert. *The Human Use of Human Beings: Cybernetics and Society*. Boston: Houghton Mifflin Co., 1954.

Williams, John. "Distant Intimacy: Space, Drones, and Just War." *Ethics and International Affairs* 29, no. 1 (2015): 93–110.

Winfield, Alan. "Making an Ethical Machine." *Open Transcripts* (blog). Accessed January 5, 2022. http://opentranscripts.org/transcript/making-an-ethical-machine/.

Winfield, Alan, Christian Blum, and Wenguo Liu. "Towards an Ethical Robot: Internal Models, Consequences and Ethical Action Selection." In *Advances in Autonomous Robotics Systems. TAROS 2014. Lecture Notes in Computer Science*, edited by Michael Mistry, Aleš Leonardis, Mark Witkowski, and Chris Melhuish, 85–96. Cham, Switzerland: Springer, 2014.

World Economic Forum. *The Future of Jobs Report 2018*. Geneva: World Economic Forum, 2018. http://www3.weforum.org/docs/WEF_Future_of_Jobs_2018.pdf.

World Economic Forum. *Towards a Reskilling Revolution: A Future of Jobs for All*. Geneva: World Economic Forum, 2018. http://www3.weforum.org/docs/WEF_FOW_Reskilling_Revolution.pdf.

Zimmerli, Walther. "Deus Malignus." Paper presented at the Institut für die Wissenschaften vom Menschen, Vienna, October 2019. https://esel.at/termin/103418/walther-zimmerli-deus-malignus.

Zuboff, Shoshana. *The Age of Surveillance Capitalism: The Fight for a Human Future at the New Frontier of Power*. London: Profile Books, 2019.

FURTHER READING

Asaro, Peter. "How Just Could a Robot War Be?" In *Current Issues in Computing and Philosophy*, edited by Philip Brey, Adam Briggle, and Katinka Waelbers, 50–64. Amsterdam: IOS Press, 2008.

Benjamin, Ruha. *Race after Technology: Abolitionist Tools for the New Jim Code*. Cambridge, UK: Polity Press, 2019.

Bostrom, Nick. *Superintelligence: Paths, Dangers, Strategies*. Oxford: Oxford University Press, 2014.

Brynjolfsson, Erik, and Andrew McAfee. *The Second Machine Age: Work, Progress, and Prosperity in a Time of Brilliant Technologies*. New York: W. W. Norton and Company, 2014.

Bryson, Joanna. "Robots Should Be Slaves." In *Close Engagements with Artificial Companions: Key Social, Psychological, Ethical and Design Issues*, edited by Yorick Wilks, 63–74. Amsterdam: John Benjamins, 2010.

Carpenter, Julie. "Deus Sex Machina: Loving Robot Sex Workers and the Allure of an Insincere Kiss." In *Robot Sex: Social and Ethical Implications*, edited by John Danaher and Neil McArthur, 261–287. Cambridge, MA: MIT Press, 2017.

Coeckelbergh, Mark. *AI Ethics*. Cambridge, MA: MIT Press, 2020.

Coeckelbergh, Mark. "AI, Responsibility Attribution, and a Relational Justification of Explainability." *Science and Engineering Ethics* 26, no. 4 (2020): 2051–2068.

Coeckelbergh, Mark. *Growing Moral Relations: Critique of Moral Status Ascription*. New York: Palgrave Macmillan, 2012.

Coeckelbergh, Mark. "The Moral Standing of Machines: Towards a Relational and Non-Cartesian Moral Hermeneutics." *Philosophy and Technology* 27, no. 1 (2014): 61–77.

Danaher, John. *Automation and Utopia: Human Flourishing in a World without Work*. Cambridge, MA: Harvard University Press, 2019.

Floridi, Luciano, and J. W. Sanders. "On the Morality of Artificial Agents." *Minds and Machines* 14, no. 3 (2004): 349–379.

Funk, Michael, Johanna Seibt, and Mark Coeckelbergh. "Why Do/Should We Build Robots? Summary of a Plenary Discussion Session." In *Envisioning Robots in Society: Power, Politics, and Public Space*, edited by Mark Coeckelbergh, Janina Loh, Michael Funk, Johanna Seibt, and Marco Nørskov, 369–384. Proceedings of Robophilosophy 2018/TRANSOR 2018, University of Vienna, February 2018. Amsterdam: IOS Press, 2018.

Gunkel, David. *The Machine Question: Critical Perspectives on AI, Robots, and Ethics*. Cambridge, MA: MIT Press, 2012.

Gunkel, David. "The Other Question: Can and Should Robots Have Rights?" *Ethics and Information Technology* 20, no. 2 (2018): 87–99.

Haraway, Donna. "A Cyborg Manifesto." In *The Cybercultures Reader*, edited by David Bell and Barbara M. Kennedy, 291–324. London: Routledge, 2000.

Johnson, Deborah. "Computer Systems: Moral Entities but Not Moral Agents." *Ethics and Information Technology* 8, no. 4 (2006): 195–204.

Lin, Patrick, Keith Abney, and Ryan Jenkins, eds. *Robot Ethics 2.0: From Autonomous Cars to Artificial Intelligence*. New York: Oxford University Press, 2017.

Lin, Patrick, George Bekey, and Keith Abney, eds. *Robot Ethics: The Ethical and Social Implications of Robotics*. Cambridge, MA: MIT Press, 2012.

Matthias, Andreas. "The Responsibility Gap: Ascribing Responsibility for the Actions of Learning Automata." *Ethics and Information Technology* 6, no. 3 (2004): 175–183.

Nyholm, Sven. *Humans and Robots: Ethics, Agency, and Anthropocentrism*. London: Rowman and Littlefield, 2020.

Schwab, Klaus. *The Fourth Industrial Revolution*. New York: Crown Publishing Group, 2016.

Sharkey, Noel, and Amanda Sharkey. "The Crying Shame of Robot Nannies: An Ethical Appraisal." *Interaction Studies* 11, no. 2 (2010): 161–190.

Sparrow, Robert, and Linda Sparrow. "In the Hands of Machines? The Future of Aged Care." *Minds and Machines* 16, no. 2 (2006): 141–161.

Sullins, John. "Robots, Love, and Sex: The Ethics of Building a Love Machine." *IEEE Transactions on Affective Computing* 3, no. 4 (2012): 398–409.

van Wynsberghe, Aimee. "Designing Robots for Care: Care Centered Value-Sensitive Design." *Science and Engineering Ethics* 19, no. 2 (2013): 407–433.

Wallach, Wendell, and Colin Allen. *Moral Machines: Teaching Robots Right from Wrong*. Oxford: Oxford University Press, 2009.

INDEX

MARK COECKELBERGH is Professor of Philosophy of Media and Technology at the University of Vienna. He is the author of *New Romantic Cyborgs: Romanticism, Information Technology, and the End of the Machine*, *AI Ethics* (both published by the MIT Press), *Introduction to Philosophy of Technology*, and other books.